73

DESIGNER'S GUIDE TO TESTABLE ASIC DEVICES

DESIGNER'S GUIDE TO TESTABLE ASIC DEVICES

Wayne Maurice Needham

VNR VAN NOSTRAND REINHOLD _____ New York

Library of Congress Catalog Number 90-38645
ISBN 0-442-00221-1

Printed in the United States of America

Van Nostrand Reinhold
115 Fifth Avenue
New York, New York 10003

Chapman and Hall
2-6 Boundary Row
London,SE7 8HN,England

Thomas Nelson Australia
102 Dodds Street
South Melbourne 3205, Victoria, Australia

Nelson Canada
1120 Birchmount Road
Scarborough, Ontario M1K 5G4, Canada

16 15 14 13 12 11 10 9 8 7 6 5 4 3 2 1

Library of Congress Cataloging-in-Publication Data

Needham, Wayne Maurice, 1949-
    Designer's guide to testable ASIC devices / Wayne Maurice Needham.
        p.   cm.
    Includes bibliographical references and index.
    ISBN 0-442-00221-1
    1. Application specific integrated circuits—Testing.
2. Application specific integrated circuits—Design and
construction.   I. Title.
TK7874.N385 1991
621.381'5—dc20

90-38645
CIP

To Pat,
Katie, and Steven
for their patience and understanding
during this project

CONTENTS

PREFACE

In 1968, when I first started in the electronics industry, I was involved in the manufacture of 4-bit TTL counters, which at that time was considered complex logic. These divide-by-10, divide-by-12, and divide-by-16 counters were considered expensive, complex, and quite difficult to test. Since the commercial test equipment industry that would support the semiconductor manufactures and users, was at that time, in its infancy, we had to design and build most of the test equipment we used ourselves.

In the early seventies, some semiconductor companies were starting to work on RAMs and ROMs as a method of integrating more transistors and more complex functions. By 1972 when I started at Intel, the 4004 model was being introduced. It had a transistor count of close to 2300 transistors—all random logic. Three other equally complex parts made up the set. We also were ramping in production the first dynamic RAM, the 1103, which had 1024 bits of RAM on a single die. For its time, these were very complex devices to test. Testing of parts for the 4000 series was also a challenge in 1972. At the time, memory testing issues were being solved by special commercial testers.

By the early eighties automated layout and design techniques along with process innovations, allowed devices in the complexity of 20,000 to 30,000 transistors to be manufactured. These models had the equivalent of 16,384 bits of RAM storage manufactured on a single device. In addition, the tools for design improved, and engineering people started using simulation extensively. Several books were written about automating design techniques, including Carver Mead's book Introduction to VLSI systems. That book, copyrighted in 1980, predicted by the late eighties that devices containing 1 million transistors would be produced. Many people laughed at the idea, and wondered what could you possibly do with a million transistors. Also during the early eighties, gate array design was becoming commercially available. Some vendors started producing standard cell libraries. The emphasis in manufacturing shifted toward testing as quality of devices became a major item in evaluating vendors. David Hodge's book on semiconductor memories quoted an AQL of 1 percent and 2 to 3 percent defective per lot.

By the mid-1980s the design of applications-specific integrated circuits (ASICs) had become quite popular. There are many horror stories about the generation of designs and the efforts necessary to get them correct. These stories included termination of employees and eventual conversion of systems to discrete logic. Estimates of a poor 30 to 50 percent success rate were common, and some designers abandoned the ASIC approach altogether. At a cost of $10,000 to $100,000 to the design, this was an expensive training experience.

Semiconductor technology advanced over the years, however, and it is now possible to implement designs of 20K to 100K gates in a standard cell or gate array device. One can complete an ASIC design in a very short time. Unfortunately, "complete" does not mean that it will work, or that it will be testable. Nonetheless, the progress in recent years has been phenomenal. It is now possible to integrate a million transistors on a single device. Carver Mead's prediction of the million-transistor milestone has proved correct. In April of 1989, Intel introduced the $860^{®}$ and the $486^{®}$ models, both containing over a million transistors. Now 5 to 10 million transistors is the next big goal.

ASIC designs of 100,000 gates, or 400,000 transistors, although not common, are being put together today. These units will become quite popular in the next few years. Great progress has been made in the areas of device design, process capability, and testing, over the past 20 years. An individual designer can now complete a very complex design using ASIC technology in a very short time. The design can be sampled in very complex high-pin-count packages and mass-produced in volume production in less than a year.

In evaluating the successes and failures of ASIC designs, one is constantly aware of the degree of effort, energy, and ingenuity of design techniques put into the testability of a device. This book will focus on these testing techniques, tracing the entire process from selection of vendors, tools, and testers to volume production. It will cover the practical aspects of testing and how self-testing can be built into devices or LSSD or direct access: What are some of the standards that are commercially available for the testing of devices? How does a semiconductor vendor take vectors generated for testing from a customer and implement them into a production testing program? How do you recover from errors? What aspects of the design of a device can impact the ability to test and to detect and isolate errors in the design?

Unfortunately, if the system designer does not pay careful attention to the manufacturing aspects of the device, specifically testing, because of the complexity of the part, the designer might end up with something that is unusable. To prevent that from happening, one must tie together the aspects of design, testing, and processing at the earliest-possible stages. Preferably, this is done during the planning stage at the system partitioning level. This will ensure that nodes are observable and accessible and that the right kind of data patterns and techniques are used to test the device.

Today, since an estimated 40,000 designs per year are being done using ASIC technology, you can easily imagine the wide variation in the design and testing capability for these devices. Parts that are designed by inexperienced ASIC designers can be made testable. This book will try to convince you to implement testing capability within the device during the initial design. This should prevent those long frustrating hours of debugging that are necessary to ensure that the device functions properly. Hopefully, this will also prevent you from pulling your hair out and cursing at the sticky test problem. What's more, the designing of an ASIC device will become a rewarding experience rather than one that you would not want to relive.

This book is meant as a guide for first time ASIC designers to help them understand some of the nomenclature and aspects of design for testing. It will cover aspects of logic circuit design that you were probably not taught in your college classes. Moreover, some aspects of the book may even prove helpful to an experienced ASIC designer. For instance the book provides information on the tradeoffs that vendors may discuss relative to die size or certain techniques that you may be able to employ relative to implementation of a device.

Since this text is intended specifically as a guide for ASIC users for implementation of testability, many tradeoffs were made in the generation of the book. The first goal was to avoid making the book a medium of advertisement. The total number of references within the book to individual companies or to specific software, hardware, or tools are at a minimum. Most of the discussions in the book are generic: They include practical applications and implementations of techniques used for testing of integrated circuits.

The achievement of a well-executed ASIC design is a very rewarding experience. The tools and process capability available today allow very complex systems to be fully implemented on a single integrated circuit. This process can take a relatively short time. Many of the tools available today including logic synthesis, block device design techniques, and automated testing techniques improve the probability for success far above that of the mid-1980s.

The present semiconductor market is estimated to be over $40 billion annually. The ASIC portion of that—$12 billion—although relatively small, is growing at the fastest rate of any market segment. Potential ASIC designers have available to them more than 200 companies offering tools for gate array, standard cell, and full custom design. Processes include bipolar and ECL, both metal gate or silicon gate MOS technologies, and gallium arsenide. The designer has the choice of working with elements as simple as basic transistors all the way to complex microprocessor functions. These designs are found in some manufacturer's libraries. There are also analog components, analog libraries, digital functions including bit slice processors, RISC processors, and micro-processor peripheral functions, along with the common NAND, NOR, AND, and OR logic gates and flip-flops.

Probably close to 20,000 designs are being generated today. This number is expected to increase dramatically over the next few years, as it will be the major portion of system design by the mid-1990s. Although the book references standard cell and gate array CMOS ASIC designs, the techniques, questions, and processes described here apply to full-custom, ECL, and other methods of ASIC design.

If you are one of the lucky people to be involved in implementing your first designs, you will find this book a valuable reference. Pay close attention to testing and you will have good luck with your *testable* ASIC device.

Acknowledgments

I would like to acknowledge the following people for their contribution to the book. Willy Agatstein and Don Nelson for the technical review of the preliminary manuscript. Nancy Ballinger and Patricia Needham for the editing of the manuscript.

Mohammad Rehman of the Signal Processing Group in Chandler for technical generation of the various figures in the book. Glen Koscal and John Heacox of Amkor Electronics for the detailed packaging information. The IEEE, Prentice-Hall, *ASIC Technology and News,* ICE, and Intel for release of previously printed material. Dr. John M. Acken for the detailed information and data on fault grading.

Finally, to Alan Goodman, for getting it all started.

1

ASIC PROBLEMS OF TODAY—OR THE COST OF DOING IT WRONG

The world electronics industry has undergone dramatic change in the past decade. First of all, there has been a globalization of the industry. Competitors may now be located halfway around the world rather than just in the building next door. Another aspect of the change is that the emphasis is now on cost, quality, and volume, and this was not so as recently as ten years ago. For companies involved in the design of systems there is also a change in the level of integration. Devices of greater and greater complexity are now commercially available, and many corporations feel the need to differentiate themselves from their competitors by developing custom or semicustom devices.

The ability to create these custom devices, or applications-specific integrated circuits (ASICs), has improved significantly over the past ten years. With this shift to applications-specific integrated circuits, there is a corresponding shift in the responsibilities for the quality and manufacturability of the devices. The designing of semiconductors is a process that requires well-thought-out access to internal nodes to ensure that the devices are testable. The focus then of the customer must shift to the following considerations:

- How do you get the device to be manufactured in a high volume with a minimal amount of production effort?

- How can the device be manufactured to achieve such consistently high quality that it is competitive in the market with the increasing expectations of your customers?

- Finally, how can all this be accomplished at a cost-effective price?

These three questions are addressed in this book, and the answers are interrelated. Obviously, the perspective taken during this book will be the one related to testability and testing as it impacts these three facets of device production. It is the premise of this book that without testability and high-quality test circuits added to integrated circuits, none of these questions can be addressed. Conversely, if not addressed, the device will not be successfully introduced into production. It is estimated that only 20 percent to 30 percent of the ASIC designs done today go into production. This low percentage is often related to the device not working correctly or not being testable.

Several factors will influence correct operation of the device; first is the proper design for the particular system use. Certain design techniques will help in the definition of the appropriate logic for system use, and there are many good books on this topic; this, however, is not the subject of this book. A second aspect of devices not going into production is the fact that many of them are not testable. These devices may function correctly some of the time, but on average, they do not always meet specification. The specification may be ac or dc, or the functionality of the part may be nonconforming.

ASPECTS OF SUCCESSFUL ASIC

Three basic elements must come together to make a successful integrated circuit: vendor manufacturing capability, including processing and packaging; the logic and design of the integrated circuit, including schematic capture, library selection, synthesis, and block selection and simulation; and testing, including testability test capability and the tester.

Most engineering people involved in the design of ASIC devices are familiar with the tradeoffs between gate arrays, standard cells, and perhaps full custom. They are also familiar with the process selection from one vendor to another. The aspect of test capability and testability is often overlooked by designers about to embark upon an ASIC design. This was probably acceptable in the past when the typical design consisted of only a few hundred to a few thousand gates. Some of these designs were implemented and then turned over to a vendor or test engineer to brute-force a test program into production. Often the test engineer had to do this without any understanding of the part and without any testability circuits added to the part.

As design complexity increases (the technology is now capable of doing 20,000- to 100,000-gate ASIC designs), this attitude about testing must change so that designers may be successful in the design and manufacturing of ASIC devices. Circuits must be put together with testability incorporated into the initial design process.

Doing a design can be very rewarding. There are many good tools and advanced processes available today that will help in the process. It is possible using blocks, logic synthesis, and an engineering workstation to complete the logic design in a very short period of time. In addition, the newer design techniques for testing help in the debugging and testing of the device. Reward in designing is often related to having it manufactured. Most engineers get some level of satisfaction out of seeing their design transfer into volume production. The transfer will not take place, however, if the design is untestable or non-manufacturable. Designs generated without testability in mind may run significantly longer to test, debug, and place into production, and sometimes they do not work out at all. When compared with testable designs, the tradeoff may be one of success or failure.

ASIC DESIGN COMPARED TO STANDARD SYSTEM DESIGN

The semiconductor business, specifically ASIC, poses a different set of problems for the system designer. Nodes are not nearly as accessible as they would be on a PC board. One cannot use an oscilloscope, logic analyzer, or other kind of in-circuit test capability for access to the internal nodes of the part. Nor can circuits be patched by common-circuit board techniques.

Although it is possible to go in and contact individual lines and measure timing and voltage parameters, this is not usually an option. Contact with lines is virtually impossible to achieve without the expensive equipment available to a large semiconductor manufacturer, who would routinely access and probe internal nodes for debugging. For an ASIC device, that may cost between $20,000 and $100,000 to design, buying the equipment needed for microprobing or electron-microscope-voltage contrast testing of internal nodes is too expensive. It is also so time-consuming that it is probably not worthwhile to use as a fallback position if testability were not included.

Thus the emphasis in the design process must initially be oriented toward preventing these kinds of problems. This is done by designing access to the circuit to allow debugging and even though not able to make actual contact with the intermediate nodes. Such a design eliminates the need for an oscilloscope or logic analyzer to probe the nodes. This important goal of the design process must be thoroughly understood before implementation of any device. Once testability is implemented into the part during the initial design phases, debugging of the device and checking its functionality and logic capability become relatively easy tasks. This book will cover many vital issues in the practical approach to design for testing.

Perhaps the biggest difference between ASIC design and system design using standard components is the ability to access, test, and debug the device. An oscilloscope and logic analyzer are good tools for the verification of logic blocks and functionality in a system. However, the use of these tools in debugging an integrated circuit is very limited. Although it is possible to contact individual lines and measure timing and voltage parameters, it is almost impossible to do so without very expensive and complex equipment.

Therefore the design process must be more oriented toward testing. Proper circuit design allows test-related problems to be debugged and verified. It allows debugging without having access to intermediate nodes, which minimizes the need for an oscilloscope or logic analyzer to probe the nodes. These tools are still good for verification of the test set, but only as a supplement to the tester. This different perspective in the design process is one item that must be comprehended before implementation of a ASIC design.

Looking at the three areas of success from the ASIC perspective—design, test, and process—is the focus of this text. Chapter 2 will focus on tester selection. Some of the nomenclature of tester manufacturers will be discussed along with the impact of testing on the design process. Chapter 3 will look at vendor selection. Design will be covered in Chapters 5 and 6, but only related to testability. Many other excellent books are available on the design of logic.

TIME TO MARKET IMPACT

Most of the book will focus on the preparation, design, and planning of techniques and tools to ensure that a design can be debugged in a rapid and accurate fashion.

Your attention to thoroughness and accuracy in design, which is emphasized in the first eight chapters of this book, will greatly influence the ability to mass-produce your devices, as described in the final three chapters. The preparation work falls into two categories. The first is the planning of test modes and implementation of logic to ensure testability of the part, which should be done in all phases of the design starting with the earliest block diagram or even in the conceptual stages. The second item that will impact testability and time to market is the preparation and thoroughness in generation of data patterns, hardware for testing, programs for debugging of the part, and aids that can be used to analyze the functionality and performance of the device.

As stated before, this time investment is primarily recommended to ensure that the time necessary to debug the part is minimized. There are many ways to relate this investment of time to design cost, but one of the more common and noteworthy ways is to look at the relationship between time to market and profit,

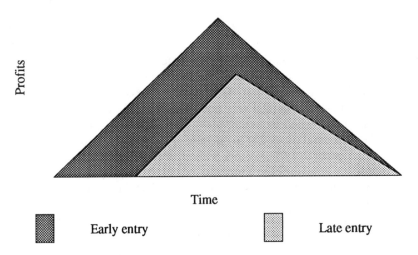

Figure 1-1. Time to market impact.

as shown in Figure 1-1. This figure depicts the delays in the initial start and ramp in production of the device and the overall input to product, profit, and market penetration.

Of course, this assumes that this device is going into an application where there is competition. In some cases ASICs are being designed for a single system with no competition, and the figure may not hold in that case. Although in the case with no competition the design of the system may be done as a cost reduction, some net costs would drop instead of overall market share gaining if the design were done and implemented rapidly. Accurate design for testability is an attempt at solving this total tradeoff of design complexity and throughput time.

COST OF TESTING

There is a cost associated with testing of devices owing to time on the test system, time to generate the patterns, incremental device size to accommodate more logic for test, and possibly a need to add extra pins for testing. Each of these costs will be looked at in this chapter. Many engineering people in the past have not paid attention to the cost of testing in the design process. When the device consisted of only a few hundred gates, this may have been acceptable. But today, designs of 50,000 to 100,000 gates may be done in a relatively short period of time, and testing must be considered in the early stages of the process.

Effect of test time on cost

Test time as it relates to cost is a relatively simple calculation. Most of the discussions in the book will be talking about test times in the 1 to 3 s category. Most commercial machines used in the production of ASIC devices cost between $500,000 and $3 million. This equates to about a penny a second for depreciation alone, assuming a very good utilization. Under normal circumstances, the machine—when including plant costs, operators, and support personnel—will cost several times that to operate.

When one generates test vectors without testability, it is quite easy to consume inordinate amounts of test time. In the cost example previously stated, 2 or 3 s, even when including the loading of overhead on the test costs, would not be an excessive test cost. If that same number turned out to be 30 s of test time, the impact of 5 to 10 cents a second for testing would then become a substantial portion of the total cost of the integrated circuit. Therefore, it is important to keep the test time relatively brief.

Benefits of testing

The advantages of additional testing are really twofold. First is the aspect of enhanced yield for the device, which has a quality, reliability, and cost impact. In many cases good yield is the reason that testing is not often debated and is often recognized as needed for successful manufacture of the device.

The second reason for providing for additional test circuitry is that it provides a means of debugging the device. There is a good chance, based on industry numbers, that a new device will not work and that it will require debugging. If the device is designed for access, control, and observability, the debugging process is enhanced. Conversely, if these two items are overlooked, debugging and manufacturing can be significantly harder to accomplish, if not impossible.

Impact of die size changes on test cost

One positive aspect of ASIC design is that the incremental gates added to the logic of a device have an extremely small impact on cost. Figure 1-2 shows the relationship of incremental logic added to a 1-micrometer (μm) standard cell design. This particular cost analysis assumes that there are no limits in terms of the physical form of the device that impact cost. These limits include items such as minimum die size for the package that is selected, maximum die size for the lithography field, and the package that is selected. If the cost of the device is outside the range of linear manufacturing cost, then the cost calculations do not reflect accurately the increased test cost. If the device is smaller than the mini-

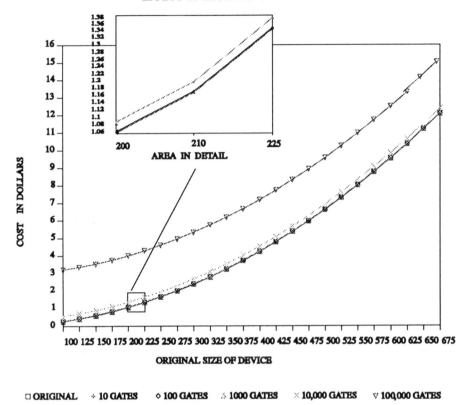

Figure 1-2. Cost of incremental gates for test.

mum required for the package, then adding incremental logic is, in essence, free. The device already needs to be expanded to fill the available bonding requirements for the device, and adding a few more gates would cost absolutely nothing. As noted in Figure 1-2, the cost of adding incremental gates is in the range of 0.001 cents per gate for a moderate size ASIC device. This same figure shows the relationship of not only 1 gate, but 10, 100, 1000, and 10,000 gates added to the original gate count and the impact on cost. Notice that the numbers are extremely small and therefore that consumption of gates for additional testing should not be a concern to the design engineer.

EFFECT OF EXTRA PINS FOR TEST

The next aspect of additional test costs for ASIC devices is related to the package pin requirements. Chapter 3, Table 3-9 shows the minimum die size requirements by package type. When converting from a 44- to a 68-pin PLCC,

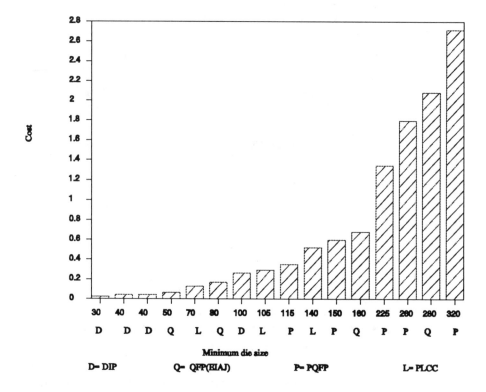

Figure 1-3. Impact on cost of higher pincount packages.

one could have the requirement of an increase in the die size by 35 mils on a side. The die cost of a 150-mm 1-μm standard cell wafer when making these transitions can be substantial. Figure 1-3 shows the minimum die sizes for various packages and the cost associated with those die sizes as they change from the 16- through 196-pin packages. One of the basic problems here is that the resolution or minimum increment in lead count for devices is finite. There is not a linear relationship of additional pins in small increments in die size, and so the conversion from 100 PQFP to 132 PQFP adds 75 mils to die size. This may be the requirement to add sufficient testability to the part.

If the design is done in such a manner that there are no additional pins that can be used for testing, the problem of adding logic becomes compounded in the eyes of the system designer. One potential solution is to multiplex not only test modes for the part, but also some of the device functionality; then use an external or a second integrated circuit to expand capability. Figure 1-4 shows an example of multiplexing to limit pin count. Notice that four pins were added for testing. This could be the P-1149.1 J-tag or other test control specification. The

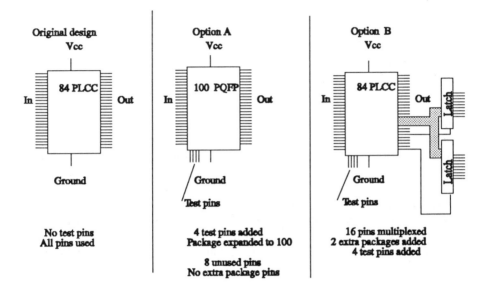

Figure 1-4. Options for saving pins for test.

ports associated with pins 10 through 20 are then multiplexed and connected to an additional latch outside of the integrated circuit. This simple multiplexing and adding of additional logic for latches can be done in various ways. The example shown here is a parallel latching of 8 bits of data for two input/output (I/O) ports, thus saving a total of 6 pins, 4 of which are used for testing. In addition, data can be serially shifted out and latched in a register, and inputs can be multiplexed for incoming data.

CHANGES IN THE DEVICE TO FACILITATE TESTING

In addition to the examples of pin count and area of the integrated circuit, there are other changes that occur in the device to facilitate testing. The foremost of these changes is that incremental logic adds incremental time in the combinatorial calculation. This must be understood early in the design process and some margin must be allowed in the process selection versus device specification. Such an allowance can make way for some logic to be added in paths to ensure observability and controllability of the nodes and states of the machine. One gate may add only a nanosecond or so to the delay in a path, but that nanosecond may be critical and must be understood during the design process.

SUMMARY

The implementation of an ASIC device entails many tradeoffs. The complexity of the design may be such that adding additional test logic is a minor problem; this is especially true in the case where large blocks are implemented that include test programs supplied by the vendors.

In addition, we must look at cost of poor quality. Although not explicitly stated in this chapter, poor testability usually means poor manufacturing yields and the inability to manufacture the device in volume. Not only does that impact cost, but untested or poorly tested devices do not work properly in the end customer's system. This is not a way of making either your present or future customers happy.

The remaining chapters of the book will look at how to overcome the problems and challenges brought up in this chapter in a manner that is least painful to the system designer. The most important point to remember is that you must pay attention to testability in the early stages of the process. Future chapters will look at how testers function relative to integrated circuits. We will examine:

- What the different processes and packages are

- How to handle some of the commercially available testing techniques, along with some examples of devices that use those techniques

- How to generate patterns for logic verification and for testing and measuring the quality of those patterns

- What happens when prototypes are generated and how are they debugged on a design verification system

- What the impact will be on long-term quality, reliability, and system cost of proper versus improper ASIC design

If by now not enough attention was paid to testability in design and you are about to receive the prototypes and have not checked hardware, software, and test program generation and incorporated the appropriate logic into the part, it is now too late to impact the design or the manufacturing process. At this point you may face long hours trying to get the device debugged—and you may not succeed. Hopefully, you have not yet implementated your device, and by reviewing these chapters you may become convinced that designing for testability will greatly impact the ability to debug and manufacture the part.

2

AN OVERVIEW
OF INTEGRATED-CIRCUIT
LOGIC TESTERS

Let us first look at the type of testers that are used for integrated circuits. These machines test logic devices in both the production and verification modes. We will examine what it is possible to test and what the testers limitations are. If a designer has a good understanding of what testers can do then many of the test-related problems will not plague the debugging and production of the ASIC device.

INTEGRATED-CIRCUIT TESTERS

There are two basic types of automated test equipment (ATE) used in the semiconductor manufacturing process for the testing of logic ASIC devices. One category is the verification tester. Verification testers are targeted toward ASIC users and engineering lab application. They are most likely not used in production. These machines typically cost less than a few hundred thousand dollars for small configuration systems, and are usually limited in pattern length, timing accuracy, voltage capability, strobe placement, and throughput. Prototype testers are typically menu-driven and quite easy to operate. They require little if any formal training to use, and may be driven by user-friendly software similar to that found on a PC. Unfortunately they may take up to a few minutes to fully exercise a complex device.

The other category of ATE is the large automated test system used by manufactures of integrated circuits and by some users for incoming inspection. This is the semiconductor manufacturer's primary environment for the characterization and test of logic integrated circuits. These machines may cost in the millions of dollars. They can run millions of test vectors without pauses and can execute a complete test program for a 100,000-gate device with a million test vectors in a 200-pin package in 2 or 3 s or less.

There are other kinds of machines such as linear testers, memory testers, and mixed-signal testers for specific devices. Memory testers are typically not used for ASIC testing. Linear and mixed-signal testers are sometimes used, depending on the application of the ASIC device. The typical memory tester has the ability only to execute patterns on a device in an array form; this includes RAMs and ROMs. These machines do not typically have a memory used for stored-response testing. Memory testers usually have pin counts of fewer than 30 or 40 pins. Typical signals include X, Y, and Z address, along with control signals for read, write, chip select, data in, and data out.

Due to the high cost of building commercial test equipment, several compromises are made in the machines to keep the cost within a reasonable range. A few million dollars is still considered reasonable in some cases. For instance, all commercial testers run on a cycle time basis. That is, even if the logic is designed to be truly asynchronous, vectors must be partitioned into cycles. This means that if the device is designed without a fixed clock input that sets the operating frequency of the part, it still must have a repetitive cycle for testing. The simulator, or test engineer, or perhaps the software package or program itself, must translate the data patterns and partition the timing into vectors for the testing of the integrated circuit. This setup is one of the most important aspects to keep in mind while designing and simulating a device. Other compromises include drive capability and accuracy. Some testers also have limited format and clock capability. All these terms will be explained in the next few sections. The conclusion is that all devices, synchronous or not, must be tested in a synchronous manner.

STORED-RESPONSE TESTERS

The basic block diagram of all stored-response testers are very similar. Figure 2-1 shows a block diagram of a typical stored-response tester with all the necessary functions denoted for testing an integrated circuit. All the items shown here are included in every machine. They may be integrated into different functions within the machine, but the basic architecture of pin electronics drivers, timing generators, data storage, sequence controller, and test flow controller exists on all machines. The reference DUT stands for "device under test" and is the only component that is physically outside the system. The interface is simply the printed circuit board or socket adapter that connects the wiring from one pin electronics input/output line or power supply to the pin of the integrated circuit. The pin electronics card is the portion of the tester that takes patterns of ones and

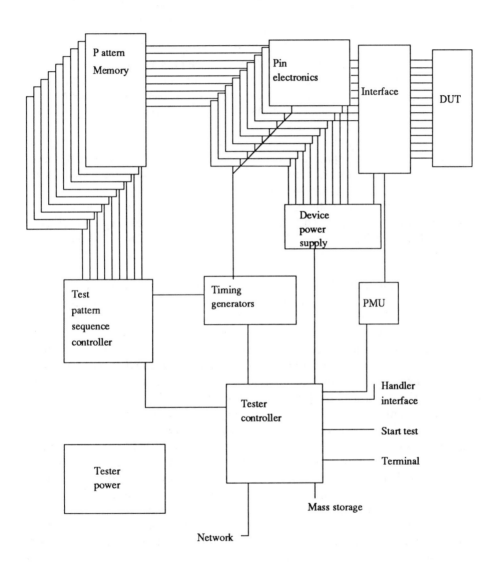

Figure 2-1. Stored-response tester.

13

zeros and turns them into V_{IL} or V_{IH} drive levels at the appropriate times. The pin electronics card also compares the output of the device for V_{OL} or V_{OH} levels with expected data and determines the pass or fail condition for the device under test. This is done on a pin-by-pin, cycle-by-cycle basis.

Pattern memory is the portion of the tester that allows storage of data patterns ordered by time to be used for forcing and sensing data. The pattern memory must contain all the information for drive data patterns and must compare data patterns. These memories may be separated into memories that control when the tester is in the drive cycle and when the device is compared. This memory may also contain format information. Formats involve the merging of timing and data patterns and will be covered in detail later.

Device power supplies are the supplies within the tester that are programmable to supply the appropriate voltage levels to power the device under test. Device power supplies also supply V_{IL}, V_{IH}, V_{OL}, and V_{OH} levels along with currents for I_{OL} and threshold voltages.

Timing generators are the portion of the tester that control the accurate placement of cycles and edges. These are referenced to an accurate crystal oscillator source. Most are digitally controlled to the 100-picosecond (ps) to 2-nanosecond (ns) increment. They place both the input driving and output compare time in the tester cycle.

The test pattern sequence controller is the section in the tester that starts and stops a test pattern. It may also load and control timing generators and complete pattern execution up to a failure point. The sequence controller may also allow looping, repeats, and "keep alive" functions. Keep alive is valuable if the patterns necessary for the testing of the device exceed the pattern memory depth of the tester. Keep alive would allow a small loop to be executed in one portion of the memory while the memory is reloaded with the data pattern for another portion of the test pattern. Keep alive is a feature useful for dynamic devices such as dynamic RAMs, and some microcontroller bus structures.

The tester controller controls the overall tester activity. This includes starting and stopping of the test, keeping track of individual test results, and interpreting start test signals either from a button, keyboard, or handler. It also controls the handlers and probers for sequencing with pass and fail information. For production machines a video display may not be available. Production machines may have a simple start test line with a pass/fail signal. Usually the tester controller has mass storage or network available that allows the storage of all the data patterns for the device under test. One million vectors on a 256-pin device is a lot of data. The tester controller also saves information for later use such as low yield analysis and characterization.

The parametric measurement unit (PMU) is a precision measurement unit that does accurate voltage and current measurements for the device under test. This allows the measurement of leakage currents in the nanoampere range and power supply currents in the milliampere range used for I_{OL} and I_{OH} testing. The PMU can measure with more bits of accuracy than the pin electronics cards can. In addition, current sensing capability is not typically available in the pin electronics cards of the tester.

Tester power supplies the appropriate supply power to the basic machine.

These blocks can be described in attributes of the machines, which will be reviewed next. The impact of these attributes on testing integrated circuits will also be evaluated including:

- Voltage accuracy, capability, and precision

- Timing accuracy, capability, and precision

- Format capability

- Features of the machine

- Control of the machine or test program

Each of these attributes will be discussed individually in the next sections. A basic understanding of these features for the intended production tester or for the verification machine that will be used should impact the design of the part. If one understands the attributes of the tester and its restrictions and keeps these attributes in mind while doing the design, the ability to test the part will be greatly enhanced. Unfortunately, many people in the past have not known or paid attention to these issues unless encouraged or forced to do so by their vendors. They may have finished with a piece of silicon that was very difficult to debug and verify on a piece of automated test equipment, or they may have ended up with a nonmanufacturable part.

VOLTAGE CAPABILITY AND ACCURACY

To test an integrated circuit, a tester needs to supply several voltages to the DUT. Commercial testers have multiple supplies that drive inputs and power supplies to the DUT. There are also sense levels for checking outputs of the part. Accuracy and capability vary dramatically from one machine to another. Typical accuracy numbers are from 1 to 20 millivolts (mV). Automated test equipment may have one, two, or more digitally controlled power supplies for the main power supply of the device; this is basically a power supply that can be turned on or off. It can be set to a voltage such as 5.25 or 3.20 volts (V) by the test program. This supply may or may not have current sensing capability to detect if the device under test is shorted or if it is drawing excess current. It may also have overcurrent protection to protect the device under test and the tester from damage if a short exists. Typically, this is used for the V_{CC} or V_{DD} supply or other power to the part.

Input and output levels may be programmed in groups of 4, 8, 16, 32, or individually by pin. The accuracy of the input and output voltages is also expressed in millivolts. Highly accurate machines are accurate to 1 or 2 mV or less. Other less expensive machines may be accurate to 10 or 20 mV. Output sense capability may be limited to a single-threshold voltage that is programmable over the device voltage range, although most machines have the capability to do both the V_{OL} and V_{OH} measurement on the same pin without resetting the pin comparators. The smaller machines may be limited to only one or two sets of output compare voltages. Bigger, more expensive machines, may have as many as one set of V_{OL}-V_{OH} pairs per pin of the machine.

The ability of the machine to measure V_{IL} on one pin or V_{OL} on another pin while holding all the other pins at a fixed value is helpful when analyzing input and output buffers. When the machine is grouped in sets of 16, the ability to measure the performance of 1 pin relative to noise, voltage and timing variations will be limited. If the tester does not have the individual pin level capability, the designer cannot tell what pin of a bus is failing a marginal V_{IL} or V_{IH} test. Table 2-1 shows some typical specifications for a device.

TIMING SYSTEMS

To test an integrated circuit, three basic types of timing generators are required for the machine. The first is cycle time. Cycle time is the basic repeating cycle of the part. Usually this is master clock or system clock. All testers need this for changes of data, timing, and formats. Some designs that are not driven by a repetitive clock will find this a difficult compromise to make. For some machines this is incremented in 10ns increments. Higher performance machines have increments of less than 1ns. This means that the tester can be set to 51-, 52-, or 53-ns cycle times for example. The beginning of each cycle is called T_0.

This repetitive pattern created by cycle time, allows the machine to take data out of the vector storage memory to force data into the device. The same cycle

Table 2-1. Typical DC specifications.

Input	Minimum value	Maximum value	Test condition
V_{IL}	-0.5 V	0.8 V	V_{CC} = 4.5 V
V_{IH}	2.0 V	V_{CC} +0.5 V	V_{CC} = 5.5 V

may be used to look at outputs of the device for correct expect data. Minimum cycle time capability on commercial machines may vary from the 5-ns range to as slow as 100 millisecond (ms). Cycle times may also be programmable or changeable from one cycle to another. An example of this is when a device needs a 50-ns cycle in cycle A and a 30-ns cycle in cycle B. This ability to alter cycle times is an excellent feature in some cases, but unfortunately it makes debugging the design difficult. If the design is bus-oriented similar to microprocessors or state time systems with fixed clocks, it may not be a feature that is useful. This variable cycle is usually found on only the largest of machines.

Inputs have their own timing generators within a tester. This is the ability to set an input to occur at a specific time within a cycle, allowing the tester to measure setup and hold time of inputs or the data bus latching time relative to clocks. With most testers the user can move clocks progressively closer and closer to an edge until the tester finds a failure point. This allows the measurement of margin relative to the specification of the device as well as the measurement of setup and hold time.

Testers may have limitations relative to input timing generators. For instance, a tester may have as few as six input timing generators that are shared between all inputs of the device. On the other hand it may be as elaborate as one timing generator programmable by cycle per pin of the machine. Obviously the cost impact of the former versus the latter will be significant. This is one of the big items that impact the cost of design verification machines versus production test machines.

Strobes are another timing generator that impacts the performance of the tester. Strobes are the time to look at or to measure the outputs of the device. They provide for the comparing of expected and actual data during or at a certain time. Strobes are categorized into two types, window strobes and edge strobes. *Window strobes* look at data from a start time to a finish time within a cycle, therefore the name "window strobe." *Edge strobes* latch data at a fixed time only. A good analogy for the two would be a D flip-flop versus an R-S latch. The D flip-flop is edge-triggered, and the R-S propagates data. Edge strobes are like D flip-flops, widow strobes are like R-S flip-flops. Again the strobes are programmed by a timing generator within the cycle time of the part. Signals must be stable during the entire window time, and for edge strobes a setup and hold time are required.

Figure 2-2 shows a typical timing diagram for a J-K flip-flop with inputs changing and the timing generators associated with them. A clock is presented on 1 pin of the J-K flip-flop. Reset, the J and the K input signals, and the output Q are represented in the figure. In addition, the timing generators associated with the clocks, reset, J input, K input, and the strobes are shown. Although this example uses four timing generators and a strobe, it is one of the simplest examples of testing. It is easy to see how the resources of the tester can be

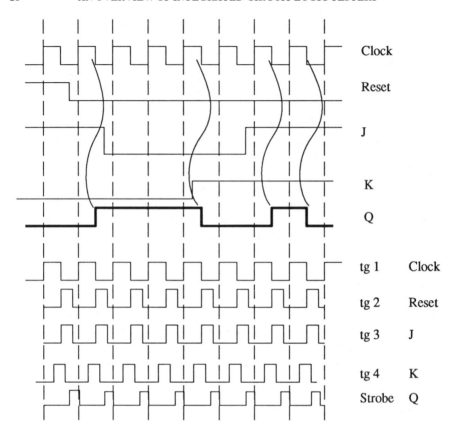

Figure 2-2. Timing and strobes for a J-K flip-flop.

consumed by testing a relatively simple device. The strobes are set to measure the output pin Q.

Voltage and timing in the pin electronics

The portion of a test system that brings together the voltage capability for input/output and the timing capability for strobes and input clocks is called the *pin electronics circuit*. Figure 2-3 shows a typical pin electronics circuit on a test system with data inputs associated for driving data, comparing data, selection of clocks for input timing generation, and a strobe for output comparison. This figure is a typical representation of a pin electronics circuit in a commercial test system. Data supplied to the driver is from the memory, timing generator, and format control of the machine. The driver controls V_{IL} and V_{IH} levels and is

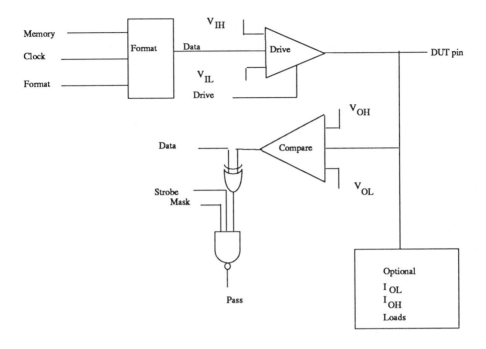

Figure 2-3. Typical pin electronics block diagram.

enabled and disabled by a drive signal. The output compare capability usually is a dual-level comparator that compares both the V_{OH} level and V_{OL} level. Data from memory is compared with the output of the device under test at the time of the strobe and then ANDed with mask data to determine pass or fail information. The I_{OL} circuitry is optional and is included on most high-performance machines.

The function of the driver is separated from the receiver or comparator section. The driver has the function of taking input levels V_{IL} and V_{IH} from the power supplies and combining them with the data pattern output from the format section. Formatting capability will be described in the next section, but for the pin electronics discussion, this allows the changing of data patterns relative to the clock supplied to the tester.

Pin electronics drivers are designed to be impedance-matched into a particular testing environment—usually 50- or 100-ohm (Ω) typical impedances. The characteristic of the driver circuit and its ability to accurately drive V_{IL} and V_{IH} levels will determine the ability to test the device for input sensitivity.

In some test systems the driver is physically disconnected in the comparison mode and reconnected during drive portions of the cycle if it is a bidirectional pin. The important parameters for the input driver are voltage swing, rise time, fall time, and minimum pulse width, which are specified for all commercial test systems.

The output capability, as shown in Figure 2-3, is a comparator that may look at one threshold voltage or at both V_{OL} and V_{OH} simultaneously. The comparator specification is probably the most critical to the test system because it is the connection of the comparator that determines the capacitive loading on the device under test. In the simulation section in Chapter 7, there will be a discussion of the effects of pin electronics drivers and their capacitive load on integrated circuits. Typical capacitive loads for commercial testers may be as low as 35 picofarads (pf) or as high as 200 pf. If in the design of an ASIC device there is no planned compensation for the tester load during simulation for test, timing parameters may be significantly different from what was specified. Ask your ASIC vendor how they recommend handling this problem.

The accuracy of the comparator is determined by the type of electronics, voltages, and impedance of the system. There is also degradation if the distance from the device under test to the pin electronics comparator and driver is more than a few inches or if there are several connections. The comparators are capable of switching within a few millivolts of the determined threshold voltage. The accuracy and speed in which the switching takes place is one of the factors that varies dramatically between small verification systems and high-performance, expensive automated testers used by commercial manufacturers.

The various outputs of the comparators are then combined with masking data and the expected comparison data to determine the pass or fail rating for the test vector on the individual pin. This is done in conjunction with the timing generator called a *strobe*.

Also included in the pin electronics section is the optional load capability. Loads may or may not be built into the tester. The function and use of the dynamic load will be discussed later.

Using voltage and timing to make shmoo plots

Using the ability of the tester to control the timing of cycles, strobes, and voltages in the small increments described here allows the user to do *shmoo plots*. Shmoo plots show the device performance, one parameter versus another on a graphic. Figure 2-4 shows a shmoo plot of V_{CC} versus minimum cycle time for a typical integrated circuit. This plot is typically called shmoo plot by most manufacturers and is available on almost all commercial test system, including verification machines. The range and increments available on the y axis for V_{CC} and the x axis for cycle time are usually user-specified. The specified operation

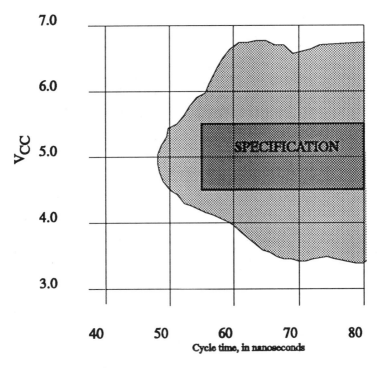

Figure 2-4. Typical shmoo plot.

region of the part is shown in the shaded area which is from 4.5 to 5.5 V, and 55 to 80 ns. The darkened area in the plot is the area where the device passes at the combinations of the V_{CC} and timing. The white area in the background is area where the device failed. Therefore, at 4V and 50 ns the device fails; but at 3.5 volts and 70 nanoseconds, the device passes.

Voltage values and timing values are varied well beyond the specification range of the device to determine pass and fail points. Each combination of voltage and timing for V_{CC} and frequency are tested and the pass condition is shown as a shaded area on the shmoo plot. This plot, when compared to the specification of the part, gives a good indication of the margin of the part relative to the specification.

Shmoo plots can be used to look at parameters other than timing and voltage characteristics of the part. Metal Oxide Semiconductor (MOS) V_{IL} and V_{IH} levels are set by ratios. TTL and emitter-coupled logic (ECL) are set by the drop of a forward-biased diode. MOS device input levels track V_{CC} changes more than bipolar devices do. Comparisons of input voltage versus V_{CC}, or of output voltage versus output currents are common. Some machines limit the number of parameters that can be shmoo-plotted in the normal operation of the machine. Shmoo plots using the PMU are available on some machines, and still other

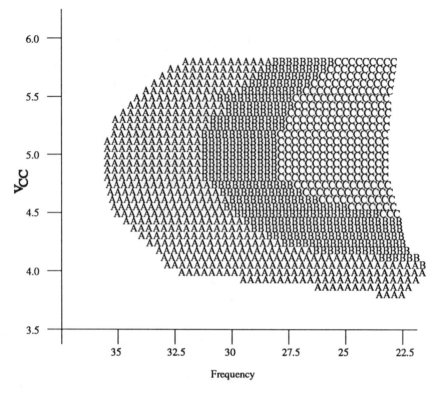

Figure 2-5. Example of a three-dimensional shmoo.

machines do not allow measured data to be used. Equations can be used to set parameters on the axis of some machines. For instance, bus timing on a microprocessor is often a function of cycle time. All the timings related to the bus cycle can be tracked if desired.

Another type of shmoo plot that is valuable is the *three-dimensional shmoo*. In this case the third dimension, z, is used for yet another data range. Examples for the use of the three-dimensional shmoo include looking at multiple parts that were tested and at the percentage that pass at any combination of V_{CC} and frequency (see Figure 2-5). This figure shows a typical three-dimensional shmoo plot, sometimes called a *three-parameter shmoo*. In this case three parts are tested for V_{CC} and frequency response and are plotted together. A is where one device passes, B is where two devices pass, and C is where all three devices pass. This plot could just as easily show different temperatures, or V_{IL} as the third parameter.

FORMAT FOR MASK COMPARE AND DRIVE CONTROL

The ability to select when the pin of the tester is comparing or not comparing, is called *mask capability*. *Masking* is simply the ability to select the cycle when the data of the device are compared with the memory data to determine pass or fail. Masking data is used on virtually every device. Typical uses are during transition when data is unstable, when outputs are in an unknown state, and during initial "power on" when storage elements are in an unknown state. Masking of data puts the output pin in a "don't care" state. Compare data is also masked during the drive portion of the cycle.

Drive control data is the ability of the tester to select which cycle, or portion of a cycle, the driver is on. Drive control simply allows the driver to be turned on and off. The data from memory are used to determine the one or zero level. Drivers are always on for an input only. In the case of an output only, they are always off. For the case of bidirectional pins, such as a data bus, the driver switches from one cycle to another to determine when the tester drives. If done correctly, this prevents contention with the device. In the case of a data bus, one pin can switch through the modes of don't care, drive one or zero, and float and compare one or zero, as shown in Figure 2-6. This figure shows the relationship of the clock input and a data bus driving different data patterns and the effect of the mask, compare, and strobe capabilities on the machine. All commercial testers have the ability to turn the driver on and off, thus allowing state changes from inputs, as shown in the first portion of the timing diagram. In the second portion, mask, strobe, and drive signals are shown. Some machines require separate drivers to be formatted for the input capability and output strobing, in which case you would consume two pin electronics drivers of the tester in order to test this bidirectional data bus. These pins are usually adjacent in the tester.

FORMAT OF SIGNAL CAPABILITY

The last feature that is important in the tester's timing system is the ability to run different formats of data patterns. Tester manufactures talk in terms of return to zero, return to one, nonreturn to zero and nonreturn to one, exclusive-or, or surround by complement formats. These patterns are described as follows:

Return to zero (RZ) data patterns always start at zero at the beginning of the tester cycle, t_0. If the data pattern for the cycle in memory is a one; during the time that the timing generator for that pin is active, the pin electronics switches to a one. When the timing generator is inactive, the pin returns to a zero. If the data pattern is a zero, there will be no change on the pin of the device. Return to one is the exact opposite of this, and the two data patterns are shown in Figure 2-7. This figure shows the timing relationship of the data pattern, clocks used for timing generation, and the signals available at the output of the pin electronics

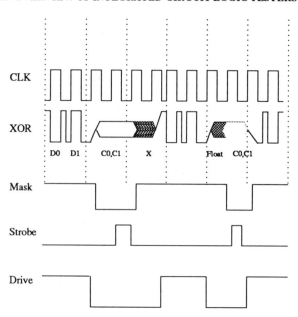

Figure 2-6. Use of mask and drive signals.

for a return-to-zero and a return-to-one format. In this figure, assume that cycle time is set at 100 ns and that the timing generator is set for 28 to 70 ns. Note the output for the RZ and R1 format.

Nonreturn to zero is a pattern that allows inputs to change to a one or zero at the time the timing generator becomes active and to stay that way through the next cycle. In this case, the second edge of the input timing generator has no

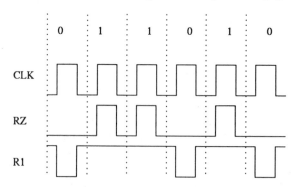

Figure 2-7. RZ and R1 data formats.

function on the pin electronics of the tester as shown in Figure 2-8. Nonreturn to zero inverted (NRZI) is the inversion of the same format, as is also shown in Figure 2-8. Also in this figure, assume that cycle time is set at 100 ns, and that the timing generator is again set for 28 to 70 ns. Note that the second edge has no effect on the data pattern.

Exclusive-or data patterns take the data and exclusively-or the data pattern with the clock generator that drives the device. If the data pattern for a particular pin is a zero, the pin electronics will drive a one to the part until the timing generator becomes active. Once the timing generator is active, the data to the part changes to a zero and stays that way until the end of the active clock period. When the clock changes to inactive, the data changes once again to a one and stays that way until the end of the cycle. If in the next cycle the data patterns switch from a zero to a one, then at t_0 the pin will switch from a one to a zero (this being the complement of the data desired for the device). Figure 2-9 shows an exclusive-or data pattern with multiple patterns running for a device. Notice the data pattern changes at t_0 and at the timing generators values. Some machines have the ability to do the t_0 switch of data patterns controlled by yet another clock generator in the system. This figure is assuming that the transition on the exclusive-or data pattern happens at t_0 from one data type to another. During the periods of time that the clock is valid, the data presented to the device is the same as what is shown on the output of the simulation data. During the periods of time when the clock is inactive, the complement of the data pattern is presented to the device. This pattern is very useful for testing data bus structures or inputs where setup and hold times are specified.

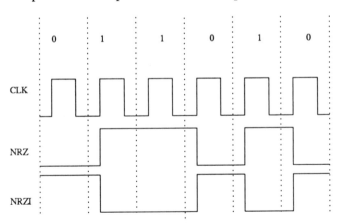

Figure 2-8. NRZ and NRZI format.

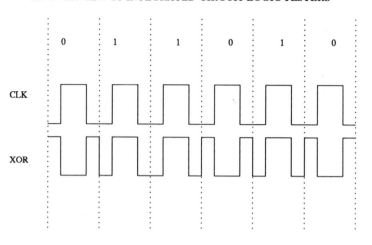

Figure 2-9. Exclusive-or format.

OTHER TESTER FORMATS

The formats listed in the previous paragraphs are only a sample of some of those available on commercial machines. Figure 2-10 shows some other available tester formats and a brief description of them. There is considerable flexibility in the industry today. Unfortunately, there is no consistent naming standard for tester formats. Note that the names used in the figure are generic and that some manufacturers use different names for different formats. For instance, the exclusive-or format, as referenced in the previous paragraphs, is often called a *surround by complement*. The basic formats needed in a design verification system for debugging and analyzing a design of an ASIC device are the ones described in the previous portion of this chapter. These additional format sets offer more flexibility and improve utilization of the tester. They allow the engineer to enhance testing of the device to ensure functionality in a wider variety of conditions, and they add flexibility for the user of the system. They also add cost to manufacture, more complex pin electronics to calibrate, and more logic in the data path.

FEATURES OF THE MACHINE

Features of the test system are basically tools or capabilities that make life as a designer or test engineer easier. This includes items like multiple-clock capability, dynamic loads, PMUs, and several other features.

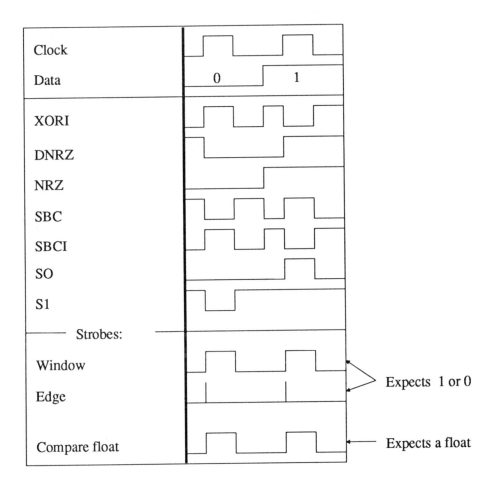

Figure 2-10. Other tester formats.

Multiple clocks

The use of multiple clocks enables the designer to generate 2, 3, 4, . . . 10 or so, usually a maximum of 16 or 32 clocks per pin per basic timing cycle (t_0). If one is designing sequential logic or a state time machine where there is a known reduction of the clock such as a divide by 3, multiple-clock capability can be of major importance in reducing vectors. This allows the compression of data patterns in the vector memory by the number of clocks per cycle. If a simulation has 30,000 vectors and the tester is limited to 16,384 vectors, the user may be able to use three or more clocks per cycle to reduce the needed memory to

10,000 vectors or less—as long as the design has three clock states per system cycle. If it has two states per vector, then only 15,000 vectors are needed. Using this capability can make a dramatic difference in the amount of vector memory needed for the tester, which is an advantage since vector memory is one of the most expensive parts of a commercial tester. The requirement is that all vectors need to have the same repeat number. If they are not consistent, then repeats should be used instead.

Figure 2-11 shows the effect of data pattern reduction by using multiple clocks per tester cycle. When systems designs are completed with a main clock frequency and divide-by circuitry to generate internal state times of the part, often there are two or three states of the clock generator representing one state of a pin. Data compression or the ability of the tester to generate two clocks for every tester cycle allows the user to use less memory for testing of the device than would be required if limited to one clock for each tester cycle. The case shown here shows the doubling of data patterns. In some device designs, divide-by-three, divide-by-four, and divide-by-six clocks relative to data pattern output capability are not uncommon.

If the tester does not have multiple clocks on a single pin, there is still a way of taking advantage of the compression aspect by using "or" logic on the interface board; this means making essentially the same function in hardware. Figure 2-12 shows the function of two pins connected together on the load board. This is an area where one must be cautious, as it would be easy to damage the drivers

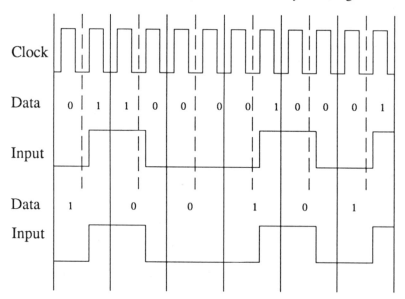

Figure 2-11. Use of two clocks per cycle.

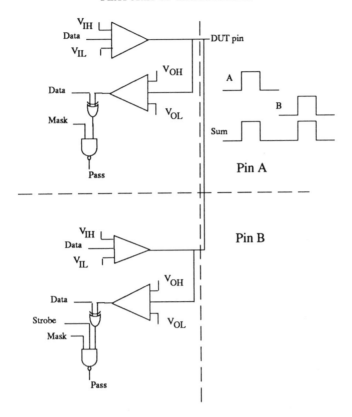

Figure 2-12. Or-combination of two drivers.

if both pins attempt to drive at the same time. The two drivers in this circuit are connected together at the load board. Driver A drives the first part of the cycle and is in a float state during the second half of the cycle. Driver B drives the second half of the cycle and is in a float state during the first part of the cycle. When wired together, these two drivers generate the data pattern, as shown in the right side of the figure. If this is used in a bidirectional pin, (data bus) only one comparator is required.

Repeats and loops

Repetition of vectors is another way to generate longer vector patterns than normally allowed by the memory of the tester. Repeat patterns allow the compression of a large portion of vector streams into small loops or single vectors. Figure 2-13 shows a typical pattern for a block of logic and the associated repeats for vectors. This figure represents repeat counts as seen in a vector

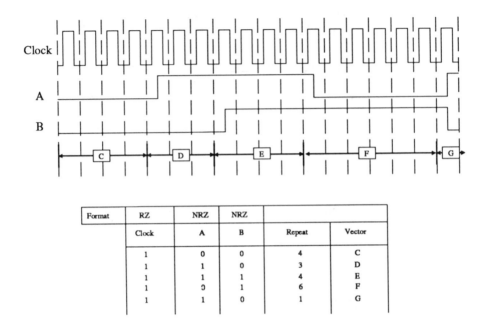

Format	RZ	NRZ	NRZ		
	Clock	A	B	Repeat	Vector
	1	0	0	4	C
	1	1	0	3	D
	1	1	1	4	E
	1	0	1	6	F
	1	1	0	1	G

Figure 2-13. Affects of vector repeats.

sequence. Clocks are generated, and the inputs A and B need to change relatively infrequently in order to exercise the device. This allows a compression of the vectors by repeating. The effect can be dramatic in some cases. Although this example shows only two vectors, it is important to note that all pins of the machine are repeated during the vector. If any pin input is changing relative to another, repeats are not acceptable. This particular example uses a nonreturn to zero format for inputs A and B and a return to zero input format for the clock.

Strobes are also affected by repeats. If a strobe is included in a cycle that repeats 7 times, then the strobe repeats 7 times.

Repeats of vectors can only be used if the full vector repeats. This means all the formats, don't cares, drive, and compare for the full vector must be identical cycle to cycle.

Repeating of vectors when using certain formats means the clock changes every cycle or every repeat cycle. Figure 2-14 shows the effect of the NRZ and RZ format with a small vector sequence, and a repeat of 2. This figure shows the relationship of the clock input to the pin electronics driver output for a return-to-zero- and a nonreturn-to-zero-formatted data pattern. With the RZ data format there are two pulses in a cycle when the vectors repeat, whereas in the case of NRZ there is only one. This can cause problems if not carefully implemented.

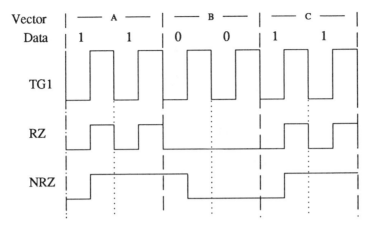

Figure 2-14. Impact of repeats on tester formats.

Dynamic loads

Dynamic load capability is the ability of the machine to load the outputs of the part to the I_{OL} and I_{OH} specifications of the part while the test program is progressing. This is a good feature in a system, and many large testers and some small testers have this capability. Dynamic loads allow the checking of the individual V_{OL} and V_{OH} levels to be done without the use of a parametric measurement unit.

Dynamic loads are made with a diode matrix, as shown in Figure 2-15. The current source at the top is for I_{OL}, and is set to the proper specification value for I_{OL}. The current source at the bottom is set to the specification for I_{OH}. The threshold voltage is set to a level between the V_{OL} and V_{OH} specifications; 1.5 V for normal TTL would be a good number, and 2.5 V for CMOS inputs (see Figure 2-15 for details).

Figure 2-15 shows the typical current load for the pin electronics of a test system. In this circuit, transistor A drives the I_{OH} of the device. At the time the transistor A is on, the device attempts to pull the output toward the V_{CC}. If the output rises above the threshold voltage which is set to perhaps 1.5 V, diode E is forward-biased and diode is F is reverse-biased. In that case the current generated by the I_{OH} current source at the bottom of the diagram is diverted from transistor A through diode E to the current source.

In the case of I_{OL} testing, transistor B is turned on, pulling the device to ground. At that point the output is lower than the threshold voltage. Diode C is turned on, diode D is back-biased, and the I_{OL} generated at the current generator at the top of the drawing is diverted through diode C to the device under test.

The settings for I_{OL} and I_{OH} are not related, and during the period of time when you are testing I_{OL} and the output of the device is at a relatively low

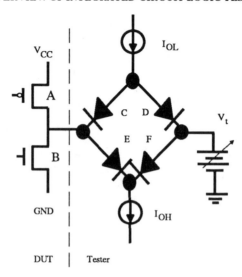

Figure 2-15. Output current load.

voltage level of, let's say, 0.4 V, the I_{OH} current source forward-biases diode F which diverts the current through the threshold voltage supply. Diode E has been back-biased and is not affecting the device performance.

Testers vary dramatically in the capability for I_{OL} and I_{OH} measurements. They may be programmable by pin, or in groups of 4, 8, 16, or so on. They may be sequenced in fixed pin groups (channels 1 to 16), or there may be several values that can be selected for each pin. In addition the accuracy of the current or voltage sources, and the ranges for these sources, should be appropriate for the type of part being tested. Make sure to check the specs for all these parameters for the part that is being designed.

Alternative I_{OL} testing

If the design verification system or test system that the engineer is using does not contain dynamic loads as described in the last section, there are alternatives. One is to take an additional power supply and hook up a resistor network, as shown in Figure 2-16. This allows the loading of I_{OL} and I_{OH} capability using an external power supply and resistor, requiring the wiring of the individual pins on the input/output board. It does not allow the switching of the loads to be on and off during tester-drive and tester-compare cycles of the part. Most testers that have dynamic load matrices built in to the pin electronics cards turn off the loads during the portion of the test that the driver is driving data patterns.

Figure 2-16 shows an alternative method of loading the device. The switch at the lower portion of the diagram would be enabled only during the portions of

I_{OL} and I_{OH} test. The value of R and V would be selected to meet the specifications of V_{OL}/I_{OL} and V_{OH}/I_{OH} for the device under test. In this particular example, the loading effect and the resistor power supply combination must be accounted for with the driver. Some drivers do not do well with resistive loads, and it may be worthwhile to use higher voltages for the input signals.

It is not a necessity to load the output during the basic functional testing of the part. The testing of the device using dynamic loads is a more exhaustive test than a basic functional test. Running data patterns or the vectors during functional test and then returning at a later time to do a parametric test is another option. Using the PMU on individual I/O pins to check input sensitivity levels and output drive and force levels is another way to accomplish the I_{OL} test. PMU testing will take considerably more time than using a vector base dynamic load test.

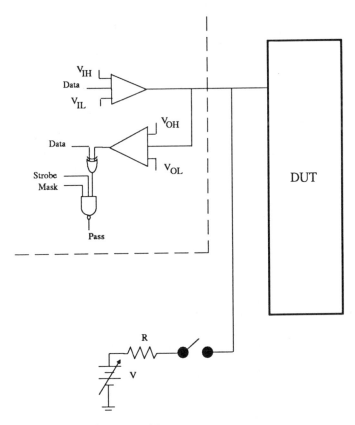

Figure 2-16. Using a resistor for current drive test.

Using the PMU is, of course, a tradeoff for debugging that must be made by the engineer. Implementing the circuitry, as shown in Figure 2-16, for a 200-pin+ device with all the additional wiring and components can be a problem. All the circuitry that is added to the load board increases the complexity and adds to the debugging problem. Remember that any wiring errors that are generated during this time may show up as testing problems, which could affect your understanding of the functionality of the part as well as the reliability of the components on the load board.

PMU

A parametric measurement unit (PMU) allows accurate measurement of dc voltage and currents. Even if the machine does have the ability to use dynamic loads for I_{OL} and I_{OH} measurements, there is probably a need for a Parametric Measurement Unit (PMU) in the machine. The PMU can allow the user to perform voltage and current measurements. For instance measure small leakage currents on input pins or make accurate measurement of I_{CC} during power testing. It also allows more accurate voltage measurement of the pins than would normally be allowed by the pin electronics in the system. The pin electronics does fast go/nogo test; the PMU is slow but does accurate dc measurements.

PMUs typically have ranges from 0 to 20 V in millivolt or microvolt increments and current ranges from 1 nanoamp to many milliamps. Some PMUs are configured only for force voltage measure current, while others allow both forcing of current or voltage and measuring of voltage or current. The capability to force a current and measure a voltage is a very useful feature in detecting the existence of junctions, during opens and shorts testing of the devices. It is also useful for checking of electro static discharge (ESD) protection devices.

CONTROL OF THE MACHINE

Control of the machine falls into two different categories. First there is the control of the pattern sequencing of the vectors, which is the execution of the ones and zero states of the test pattern. The second is the basic control of the test program, which is the setup of and forcing of the voltages and currents, starting and stopping of testing, or the evaluation of pass and fail information to determine if the part is good or bad. It is also used for binning data, data reduction, and data logging of device parameters.

Pattern sequencing control

Vector control has several aspects, the first of which is the control of sequencing of the vectors. Some testers allow looping within vector memory of short loops to exercise repeatedly the functionality of a part. Figure 2-17 shows a sequence of vectors being impressed on a part for setup of a test condition and a representation of what is happening in vector memory.

Loops are different from repeats, as repeats do only one vector. The ability to repeat individual vectors, allows pattern compression. In the case shown in Figure 2-17, one vector may be repeated multiple times during the sequence of the test pattern. This figure also shows the effect of data compression using loops. The vector sequence of 1 through 20 on the left is compressed into five different vectors—vector 1 is repeated three times; vectors 2, 3, 4, and 5 form a loop that is repeated 4 times; and vector 6 is the final vector repeated 1 time. This shows the effect of data compression by looping and repeating vectors, which is especially helpful for the purpose of a self-test capability. An example is when a self-test is started and a routine is executed until the end of the self-test capability or a result is read. Figure 2-18 shows a typical single-vector repeat sequence and what may be happening in the device during that time. This is an application where a single-vector repeat would be worthwhile. The block, in this case a 2K-bit RAM, has built-in self-test capability. A start signal and clock are supplied via external vectors (this could be part of the internal test modes), and a pass-fail line is available. The sequence for testing would be the start; the test process, which is to clock the RAM 2048 times (one clock for every state of RAM); and, finally, to read the pass-fail bit at the end. The actual repeat number would be the needed amount for all the RAM patterns.

Basic control of test system

All testers allow some kind of control of the test system flow and the tester resources. The easiest to learn and understand is a menu-driven tester. This allows the programming of the tester via a series of menus. These menus control the generation of data patterns for the tester and for the forcing of data. They also contain compare and expect data for the device under test. These menus typically do the setup of voltages and timings for the tester.

Shell- or menu-driven test programs usually limit program flexibility and are not usually used for production test programs. Testers that support a menu-driven program generator, may also have a BASIC or Fortran language for tester programming. Programming in a higher-level language such as C or Pascal is available on many machines. Usually, the language of BASIC, Fortran, Pascal, or C is a standard compiler or interpreter. Tester-specific statements such as force voltage, set format, measure current, run pattern, if fail then, etc., are

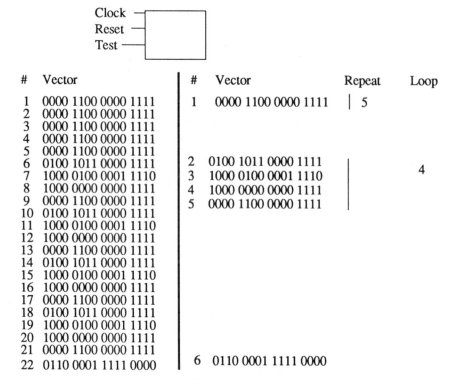

Figure 2-17. Vector loops and repeats.

added to the syntax. Programs may then be structured either as a series of calls to specific routines that set voltage, timing and execute patterns, or they may be straight line code.

Length of patterns

As mentioned several times so far, one of the major differences between large commercial test systems and verification systems is the length of the data pattern that can be run without interruption. Most equipment, even the smallest, allows for expansion of the data pattern memory. Machine pattern memory can vary from as small as 4K vectors to as large as 8 million vectors behind each pin of the tester. In addition, machines may have just a single bit representing the one or zero in the data pattern, or they may have multiple bits combining data pattern, formats, and timing information for every bit in the memory. In the first case, 4K of data patterns means 4K bits of memory. In the second case, 4K of data patterns may mean 12K or 16K bits of data behind each pin to include

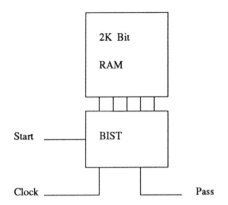

Start	Clock	Pass	Repeat
1	0	X	1
0	1	X	2048
0	0	1	1

Figure 2-18. Use of repeats for self-test.

selections of formats and timing generator information. Combining this complexity of pattern with the voracious appetite of simulators for memory needs means that vector patterns and data become quite long.

There are three items that will impact the block storage requirements of patterns and the reload time. The maximum length is related to simulation time. First is the length of the pattern memory which determines how often it must be reloaded. Next is the number of bits in the pattern memory which is controlled not only by the width of the vector (30 pins or 200 pins) but also by the number of bits representing each one and zero in the data pattern. Finally, the effective transfer rate of data from the mass storage system to the tester memory. Figure 2-19 shows this relationship of data pattern reload time from the slowest to the fastest. The highest-speed data pattern transfer rates are in the megabyte-per-second range; the lowest-speed rates may be as low as 2400 baud (Bd).

The major impact of long reload time is simply the amount of time necessary to load the patterns and loop on portions of data patterns. If the system is going to be used strictly for design verification, this may be an opportunity to read a book or manual during data pattern load time. Some of the slower systems can take 5 to 20 min to load large data patterns. This may not seem like a long time, but if you are waiting for the system to load memory to find out if the prototypes are working, this can seem like an eternity.

TESTER ACCURACY

As a final note on testers, they are all delivered to a set of specification that unfortunately are not standard in the industry. Of all the specifications, timing

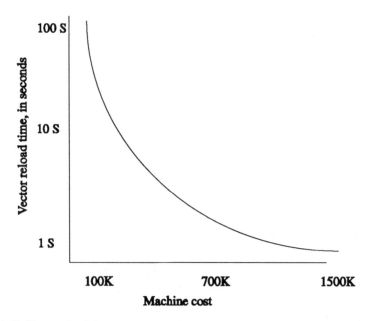

Figure 2-19. Tester reload time versus system cost.

accuracy is the most ambiguous. Most manufacturers specify timing relative to the National Bureau of Standards and can trace a standard back to NBS Pin-to-pin accuracy may depend on 1 pin in different modes, 2 pins on one tester, 2 pins on one tester with different format, or the same format on two different testers. Accuracy numbers for timing may vary 2 to 4 times depending on how it is measured. Make certain that you understand the system specification before committing to a tester.

SELECTING A TESTER

Now that most of the terms relative to testers have been explained, it will be easier to select a machine for the intended application. The largest single variable in automated test equipment is the cost of the machine. The cost may vary from as low as $100,000 to as high as several million dollars. Selecting a machine then becomes a tradeoff between cost of the machine, flexibility in testing, accuracy of the system, and the overall test time which includes reload time.

It is possible to test high pin-count parts on testers that have a lower pin count than the device you want to test, but the problems associated with doing this are substantial. Figure 2-20 shows a block of logic with a combination of inputs, outputs, and I/O's that total 68 pins. In this particular example, the minimum configuration tester would be a 40-pin machine to test the part. The minimum number of inputs plus one output on this block of logic is 38 pins. Most machines come in even increments of channel numbers. When using this configuration remember that there would have to be as many as 28 times the data patterns generated to test it. This is a ridiculous example, but it's used to illustrate the ability to test devices that presumably cannot fit on the automated tested equipment selected.

If the logic in the part is totally isolatable, one section from another, testing may be divided even more. In this case, two or more blocks may be done in different passes. Each alternative listed may require new sockets or relays to control pin multiplexing. Remember that device pin count has been growing consistently over the past several years, and all the data show that the trend will continue. Keeping this trend in mind—along with many of the other electrical parameters—will help you select the appropriate piece of automated test equipment for your purposes.

Table 2-2 illustrates a table used for the evaluation of vendors for the selection of test equipment. Parameters that are important to testing cost and throughput time are listed, and the potential of the weighing factors is added. One last note about this matrix, several items are listed which include power consumption and air conditioning requirements. For some of the larger test systems they must be contained in a room similar to a computer room with controlled atmosphere for temperature and humidity. The facility's cost for air conditioning may add a substantial amount to the expense of operating the machine. Table 2-3 lists some of the major tester and verification system manufacturers. Make sure the tester you choose and the tester your vendor uses can support the ASIC you are designing.

OTHER CONSIDERATIONS

Using the less flexible machines means more work in the design and test function. These machines can usually do a good job at verification; they are targeted for the lab debugging of devices and perform that function very well. In most cases, the use of the machines will be sporadic. It may be hard to justify using a multimillion-dollar tester for characterization and lab work. The counter argument is that debugging and characterization of the device should be done on the same machine that will be used to test the device in production. Having separate testers may cause correlation problems between debugging and production. Neither answer is always correct.

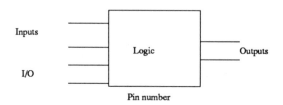

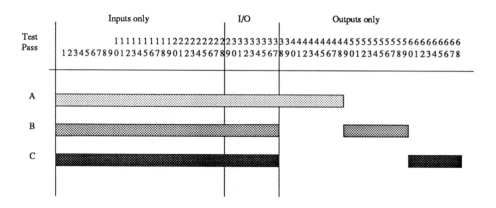

Figure 2-20. Multiple pass testing.

There are also specialized testers such as mixed-signal testers that allow the accurate testing of analog and digital functions integrated on one device. These machines are very useful if the design contains mixed signals or is primarily analog. Full analog design can be tested on these machines, but only with some difficulty and compromises in accuracy. Dedicated analog testers do a far better job.

Most of the discussion so far has been focused on digital testing of the integrated circuit. The comment here also applies to PC board testing, and the same restrictions of formats and timing also exist. All the descriptions of design verification machines mentioned in this chapter are for device verification and test.

OPTIONS FOR TESTERS

Some manufacturers sell options to the machines that allow testing of ECL, CMOS, and BICMOS with greater accuracy. Usually, this involves changes in

Table 2-2. Vendor selection matrix.

Weight	Vendor A	Vendor B	Vendor C
Pattern length			
Number formats			
Accuracy			
Speed			
Cost			
Network			
Power			
a-c Needs			
Floor space			

the design of the pin electronics to limit the voltage swing of the pin electronics driver. For instance an ECL pin electronics test head may not allow voltage swings for the full testing of CMOS inputs.

Many additional options are available to commercial testers. Typically the purchase can include hardware such as networks, PMUs, special scan memories for scan testing, special timing test capability, and mixed-signal options. In addition, some machines offer algorithmic pattern generation options for X, Y, Z addressing of memories. Separate data storage memories for ROM testing may be added. Software options may include data reduction programs, offline compilers, tester simulators, and data acquisition software.

Table 2-3. Top logic tester manufacturers

Vendor	Machine types
Advantest	P
Ando	P
ASIX	V, P
Hi-level	V
IMS	V
LTX-Trillium	P
Megatest	P
STS	P
Slumberger	P
Teradyne	P
Tektronix	P
V= Verification	P= Production

SUMMARY

The importance of adding logic for testability is often completely ignored for a variety of reasons including cost, performance, and the designer's time to implement it. In almost every case this is just rationalization. The cost of adding logic is very low, the performance impact can truly be minimized, and if you are lazy now, you should be aware that it will take even more effort to fix the problem in the future.

3

SELECTION OF PROCESSES, PACKAGES, AND VENDORS

One of the most essential aspects of ASIC devices and of the systems they are created for is the relationship between the complexity of the device and the pin count. As the number of gates increases, the number of pins required for system implementation peaks and then decreases. Several years ago a study was completed that illustrated this relationship of pin count to gate density. Although the study showed a need for increase pin count as the complexity increases, there is an inflection point where eventually all components can be integrated into a very minimal number of pins (See curve shown in Figure 3-1). This figure shows the effects of high levels of integration. As the gate complexity of the ASIC device increases, the number of pins needed for interconnection from one device to another increases. This happens until the point where integration has a tendency to include more functions and therefore requires a lower pin count. The simplest case, of course, is an integrated circuit that has two power connections, one input and one output connection for a total of 4 pins.

The smaller processes of 1 micrometer (μm) and less allow the integration of a significant number of gates. This may be a sufficient number of gates to forego the need for multiple ASICs in the design of the system. This is yet another consideration that needs to be weighed for tradeoffs of system design. Understanding this process, will help in the selection of vendors, processes, and packages. All these items will be addressed in this chapter.

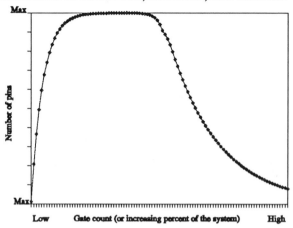

Figure 3-1. Number of pins versus gate count.

PROCESS SELECTION

Commercially available logic processes vary from greater than 3 micrometers (μm) to less than 1 (μm). Common sizes of 5-, 3-, 2-, 1.5-, 1- and 0.8- μm geometry process are in production.

The definition of the process geometry size is usually related to the transistor channel length. Measurement criteria from one vendor to another varies. When first reviewing processes, one vendor that states they have a 1 μm process, which may be slower than other vendors with a 1.5 μm process. This may be due to the measurement technique used.

The most common variation from one vendor to another relative to process measurement is $L_{effective}$ versus printed polysilicon (gate) geometries. At times drawn transistor polysilicon is also used. Figure 3-2 shows a typical MOS transistor using both measurement techniques. The differences in measurement geometries of the transistor based on $L_{effective}$ or printed polysilicon geometries is shown. Notice there is almost a 50 percent variation in the size of the process based on which measurement technique was used to measure the size of the process.

When analyzing vendor's data sheets, find out what dimension they are measuring. This will allow an apples-to-apples comparison. Other aspects that impact performance include capacitance (both junction and interconnect), oxide thickness, electron mobility, and others.

The number of functions that can be implemented on a device, although dependent on the process transistor length, is also related to several other parameters, such as cell size and metal pitch. One standard measurement for a cell size is the 2-input NAND gate. When comparing geometries, it is

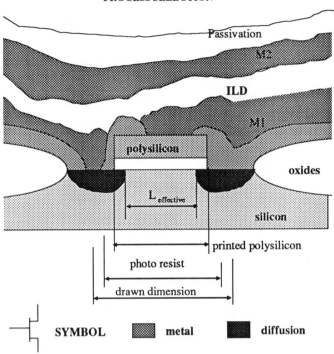

Figure 3-2. MOS transistor cross section.

worthwhile to look at the physical area in square micrometers of silicon consumed for the two-input NAND gate.

Some vendors support multiple-drive strengths, and this may cause confusion. Pick the device that has a fanout of two, and do a comparison of one library versus another with that NAND gate only. It turns out that the two-input NAND gate is still one of the most commonly used devices in most ASIC designs.

When the interconnection between gates and storage elements is primarily random in nature (meaning not a defined structure like a PLA, RAM, ROM, or data bus), other aspects dominated density. In this case, the primary factor affecting density is the interconnect of Metal 1 and Metal 2. Again, when sorting through the complexity of vendor data, common items that are referenced are the width and space of Metal 1 and Metal 2. Those two parameters may look relatively good, but the most important factor is the metal pitch.

Pitch is defined as the combination of a metal line and a space placed as closely together as possible for the lithography used. Even with this definition of metal pitch, there are variations in design rules based on contact and via alignment. Metal pitch can be defined many different ways. Depending on the

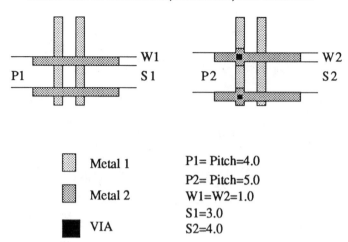

Figure 3-3. Metal pitch.

process, dog-bone geometries, which are the widening of metal around contacts and vias may be necessary. This will appear as a different metal pitch than on minimum distance parallel lines.

Figure 3-3 shows a metal-one and metal-two structure with and without adjacent contacts and vias. Again, notice a sizable difference for interconnect that can be placed in an area depending on the measurement technique used.

These two primary metrics, the size of the transistor and the metal pitch, drive the designer to the all-important question: How many devices or gates can be integrated on a single integrated circuit of given size for the type of process? Of course this measure is relative to transistor and gates only. Large blocks like microprocessors, RAMs, PLAs, and ROMs will increase density if used effectively.

After one knows what the basic parameters are, such as transistor performance, physical layout of transistors, and relative density of the process, the decision process can proceed. The next item for evaluation is the cost of the wafers or the cost of the individual devices being generated. Items that impact the cost include the number of masking steps, the cleanliness of the fabrication area, the photolithography limits, and the types and complexity of the process equipment used. The variation in equipment type and cost between a 3-μm and a one or sub-one μm process is dramatic. A 3-μm process will easily use projection print equipment for full wafer lithography.

A 1-μm process probably uses reduction steppers. To give an idea of this simple tradeoff in processing capability, Table 3-1 shows the relative cost and throughput of a projection printer versus a reduction stepper. This table shows the number of wafers per hour through the masking steps of a 3-μm and a 1-μm

Table 3-1. Stepper cost and throuput.

Equipment	Cost	Wafers per hour	Relative cost per wafer
Projection printer	300,000	60	1
Stepper	800,000	20	8

process. When the cost of the lithography equipment is included, it is easy to see why the smaller geometry processes are more expensive.

In addition, a 3-μm process is significantly simpler than a 1-μm process to manufacture. Table 3-2 shows the cost estimates from ICE between these CMOS processes. Also included is the cost for 100- and 150-mm wafers.

If you compare these items, it is easy to come to the understanding of why the highly advanced processes are so much more expensive.

Physical gate count or area consumed per transistor and speed is significantly better for a 1-μm process than for a 3-μm process. The type of design that is being done may not call for the advanced process with its associated added cost. Because the die size decreases and the cost of the equipment increases, the net impact is not as important as it may appear for the total device cost. In fact, in some cases it may be cheaper to run devices with the smaller geometry process than it is on the larger geometry process. Figure 3-4 shows the cost curves for two different processes versus gate count. These curves represents the cost of the device based on the number of devices implemented for a 3-mm and 1-mm process. The important factor here that causes the lines to cross is the number of pads or area of the integrated circuit required to be used for bonding. This

Table 3-2. ICE wafer cost estimates.

Cost of wafers processed		
Wafer Diameter	Process	Wafer cost
100 mm	3 μm	$100
150 mm	3 μm	$150
100 mm	1 μm	$400
150 mm	1 μm	$600

Courtesy of ICE

example, assumes that the maximum number of bond pads are installed on the device. The number of bond pads would vary according to the number of gates available.

Design and library selection

Vendor's libraries can vary from as simple as less than 100 AND, OR, NAND, NOR gates to moderate-complexity MSI functions to 500-plus total cells including most TTL functions. These functions may be committed in what is called *soft macros* or *hard macros*. Soft macros involve the layout of moderate to large logic functions using basic building blocks such as NAND and NOR gates, which are allowed to be placed randomly in the layout. Hard macros involve a layout of the function using hand-packed transistors or blocks fixed together to get the maximum density. Hard macros also have fixed timing, whereas soft macros, because they vary with layout, vary timing.

The cost and die size impact may be substantial for any particular cell when looking at the variation of a complex function between a soft and a hard macro. One has to remember that what is important is the number of instances within the device. Typically, the larger functions such as multiplexors are used less often than the two-input NAND gate, for instance. A small variation between two-input NAND gates may have a larger variation on device total cost than what seem to be a sizable impact die size from the larger MSI functions.

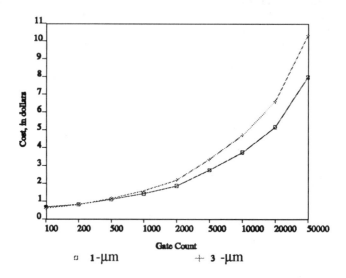

Figure 3-4. Comparison of 3-μm and 1-μm process cost.

In addition, note that for gate arrays, virtually all functions are made from two-input NAND gates, and there should be minimal impact for MSI blocks. Figure 3-5 shows the tradeoff in design techniques in physical area for a flip-flop made from a soft and a hand-packed technique. Notice that the physical die area deltas look rather large, but remember that most designs will not use them as often as NAND gates. This figure shows two implementations of a flip-flop, one hard macro and one being a soft macro. Total transistor count is 28 versus 46.

Estimation of gates and speed

Based on the kind of designs that is selected, speed, power, and density may have a great impact on how the design is to be implemented. Although a design can compensate for the basic speed of the process via optimizing logic for speed, it is not possible to speed up a design beyond the basic elements in the library. If the device that is being designed is not going to be speed-limited on the process, almost any of these processes will do the job. For instance, if the design selected is a 1- or 2- megahertz (Mhz) design and if it is marginal, the design may be done twice. The first with the logic as designed and again with logic optimized for speed on a slower, less expensive process using more gates.

Flip-flop Comparision

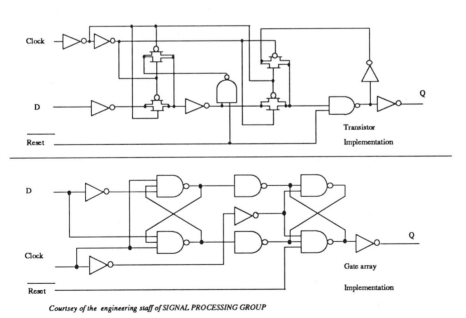

Courtsey of the engineering staff of SIGNAL PROCESSING GROUP

Figure 3-5. D flip-flop comparison.

The next step is to do a rough approximation of gate count. Several techniques are available including schematic entry, logic estimation checkers, and counting of logic elements that will give an approximate number of gates to be implemented in the ASIC design. Using a schematic entry package on a workstation gives the best results.

Once a method is used to obtain the number of gates and the interconnect factor from the vendor is known, one can go through a relatively simple calculation to find out the die size of the proposed final design on the processes that are being evaluated. Based on the lithography differences described in the first section, die sizes also may be limited by many concerns. First, the lithography type may be projection printing, 1X or 5X reduction steppers which limit maximum die size. Secondly, maximum allowable die sizes for packages vary based on the pin count of the package and the rules used by the vendor for the assembly process. Finally the same package rules limit minimum die size based on bond wire length. Table 3-3 shows minimum and maximum allowable die size per package and the approximate gate count for two different processes. Assuming that there was already a decision on the logic functionality for the device, there may not be a choice in the package. The design may have to use a higher performance process to contain the design. It is worthwhile when wondering if the design is close to the boundary of capability to go to the next most advanced process.

Most of the rest of this book will emphasize how to add logic to ensure that the device is testable. If there is no room to allow the design to grow and add the logic, there may not be space available to add all the desired test circuits. These circuits may dramatically impact the ability to debug, test, and manufacture the device. Typically, circuit additions are in the 5 to 20 percent range.

Gate arrays and standard cells

Choosing between a gate array and a standard cell design implementation is another area that will impact the design. Some of the basic tradeoffs are flexibility, throughput time, final production cost, and nonrecurring engineering (NRE) charges. Table 3-4 shows the relative tradeoff of cost, throughput time, die size, NRE, and volume for different custom design techniques including standard cell, gate arrays, full custom, and programmable devices. Notice that these characteristics vary dramatically. Based on the requirements for long-term volume the engineer may or may not need to or want to use standard cells or gate arrays. Conversion between a gate array and a standard cell may require the conversion of vendors, or the selection of one vendor and capability over another. Figure 3-6 shows gate array and standard cell cost per unit data relative to volume. This figure includes only gate arrays and standard cells. When including the cost of NRE for a gate array and a standard cell the figures change.

Table 3-3.   Gate count for various packages.

PACKAGE TYPE	LEAD COUNT	MINIMUM DIE SIZE, MILS	MAXIMUM DIE SIZE, MILS	3 -μm MIN-MAX GATE COUNT	1 -μm MIN-MAX GATE COUNT
PDIP	16	030 X 030	145 X 500	225-18125	2025-163125
(300)	24	040 X 040	150 X 500	400-18750	3600-168750
PDIP	24	040 X 040	390 X 600	400-58500	3600-526500
(600)	40	070 X 070	390 X 650	1225-63375	11025-570375
	48	100 X 100	390 X 650	2500-63375	22500-570375
PLCC	44	070 X 070	400 X 400	1225-40000	11025-360000
	68	105 X 105	700 X 700	2756-122500	24806-1102500
	84	140 X 140	900 X 900	4900-202500	44100-1822500
PQFP	84	115 X 115	400 X 400	3306-40000	29756-360000
(Bumpered)	100	150 X 150	500 X 500	5625-62500	50625-562500
	132	225 X 225	700 X 700	12656-122500	113906-1102500
	164	260 X 260	900 X 900	16900-202500	152100-1822500
	196	320 X 320	1000 X 1000	25600-250000	230400-2250000
QFP	44	050 X 050	200 X 200	625-10000	5625-90000
(EIAJ)	64	080 X 080	350 X 350	1600-30625	14400-275625
	100	160 X 160	350 X 580	6400-50750	57600-456750
	160	280 X 280	850 X 850	19600-180625	176400-1625625

Package estimates courtsey of Glen Koscal, Amkor Electroincs, and
John Heacox, Amkor Electronics.

These cost curves are shown in Figure 3-6; notice how they cross. Low volumes are cheaper to manufacture in a gate array. High volumes are lower in cost in a standard cell.

Some vendors offer the capability to develop gate array designs and rapidly convert them into standard cell designs. This takes advantage of the low cost of using standard cells for high-volume production. It also takes advantage of the fast throughput time of gate arrays for prototypes. If the volume meets a certain minimum criteria, this may be very well worth investigating. Remember when converting a design that cost will include paying for two sets of nonrecurring engineering charges. The production during the first portion of the product life cycle will probably be gate arrays. There may be device performance variation when converting from a standard cell to a gate array. That conversion needs to be analyzed and executed carefully.

VENDOR PERFORMANCE

It is estimated that there are over 200 semiconductor companies in the standard cell and gate array business. Table 3-5, courtesy of ASIC Technology and News,

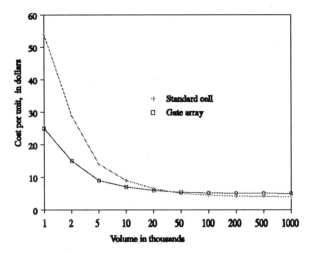

Figure 3-6. Standard cell versus gate array cost.

shows the top 30 vendors. Notice that no single large vendor truly dominates the field. The sales by vendors are often relatively small in comparison to standard product sales of the same corporation. For the 200 companies in the ASIC business, certain financial factors impact their ability to serve the customer. Depending on the design, it may be worthwhile to look at the size, profitability, and track record of the companies that are being evaluated. Most companies can provide recommendations from satisfied customers.

Table 3-4. Design and technology comparison.

	EPLD	GATE ARRAY	STANDARD CELL	FULL CUSTOM
Design cost (NRE)	5K	20K	60K	1500K
Design time	2 weeks	2 months	5 months	15 months
Die size	400	300	250	200
Flexability	Lowest	Moderate	High	Highest
Production cost for:				
1K	GOOD	POOR	POOR	POOR
10K	FAIR	GOOD	POOR	POOR
100K	POOR	FAIR	GOOD	POOR
1M	POOR	POOR	FAIR	FAIR
10M	POOR	POOR	POOR	GOOD

Certain companies are specialists in the ASIC area, and for others ASIC is a secondary function within the framework of a larger standard product organization. Throughput time and cost will vary from vendor to vendor. Items such as nonrecurring engineering charges may be waived to get business in some cases.

Note that over the past several years many companies have exited the gate array and standard cell market. This may further impact the ability to execute designs. It would be worthwhile to ask the vendor how successful they have been in this particular business segment. You can then use this information to determine whether you think they will stay in business. Ask the question: Will they continue to support the design long-term?

Vendor test aids

Some vendors have aids for the generation of test capability. If the vendor does not specifically support test aids, most of the same techniques can be implemented in an ASIC design using standard logic elements in most cases. The difference is ease of implementation. Some of these techniques will be briefly described here. A more detailed explanation and examples exist in the future chapters.

Direct-access testing is the use of multiplexors on inputs and multiplexors in front of the outputs to select internal nodes for observation and control during test. This technique allows for checking of the nodes that would normally not be accessible to the outside of the package. Examples of use of a direct-access test scheme would include control and observation of internal state flip-flops, internal ROMs, PLAs for instruction decoding, RAMs and large logic blocks. Any other large block that needs complex patterns that is not connected to the package pins directly is a candidate for direct-access test.

Self-testing uses the design of test logic to check a block of circuitry. This technique usually requires adding more logic than the direct-access scheme but takes over the test function altogether. Due to the larger test logic overhead, blocks such as large RAMs, ROMs, PLAs and multipliers are candidates for self test. Self-testing uses extra logic to generate patterns, compare data, and determine the correct functionality of the block. These patterns check the data, and determine the validity of the logic. Outputs can be as simple as a pass/fail signal on one output of a gate, to a signature, or multibit polynomial representing the states and sequence of the device. Any returned signal must be recognized by the tester and test program as a fail or a pass signal. Bilbo is a form of self-testing.

The level-sensitive scan design (LSSD), or scan test, is a technique that replaces the normal latches (D, R-S, J-K) with dual-use scan latches. In normal operation, the latches act as normal storage elements for the device. In the test

Table 3-5. Top thirty ASIC manufacturers.

Rank	Name	Sales	Percent market share
1	Fujitsu	488.0	9.9
2	NEC	455.0	9.3
3	LSI Logic	367.0	7.5
4	Toshiba	360.0	7.3
5	AMD	349.8	7.1
6	AT&T	268.0	5.5
7	Texas Instruments	217.3	4.4
8	Hitachi	203.0	4.1
9	National	173.0	3.5
10	Motorola	140.0	2.8
11	VLSI Technology	125.0	2.5
12	Oki	109.0	2.5
13	Plessy	102.0	2.1
14	NCR	92.0	2.1
15	Seiko/SMOS	90.0	1.8
16	Honeywell	79.0	1.6
17	Philips/Signetics	76.6	1.6
18	Ge Solid State (RCA)	55.0	1.1
19	Matsushita	47.0	1.0
20	Mitsubishi	43.0	0.9
21	Harris	42.8	0.9
22	Hughes	42.2	0.9
23	SGS-Thompson	42.1	0.9
24	Siemens, AG	41.0	0.8
25	Intel	41.0	0.8
26	Raytheon	40.0	0.8
27	Altera	37.0	0.8
28	Cypress	34.0	0.7
29	IMP	34.0	0.7
30	VTC, Inc.	31.5	0.6

Coutresy of ASIC Technology and News

mode, the latches are connected one to another to form a shift register that can easily move data into and out of the circuit.

The shifting of data in allows the easy control of all the combinatorial elements within the device. This can be used for the setting up of test conditions for blocks of logic or for actual testing if needed. At the same time that data is being shifted in, the end of the shift resistor shifts data out. Data out is compared to pregenerated patterns for pass or fail determination. Through the use of scan, the logic can be checked for correct functionality.

Depending on the ratio of latches to combinatorial logic, scan latches can add a considerable amount of area to the design. This is one reason that the technique is not used for all designs. True scan, or LSSD, assumes that all storage elements within the device are dual-mode scan latches. There are partial-scan techniques that allow the use of scan latches to check the boundaries of blocks and the interconnections between blocks. For designs that are partitioned into smaller logic blocks, this is a good compromise. Partial scan minimizes the cost impact of scan design.

JTAG, or IEEE P1149.1, is a method for the checking of traces used for interconnection on a printed circuit board. This ensures that the output of one pin is connected to the input of the next device on a PC board. On some of the higher-performance, more complex surface-mount packages, a visual inspection of the solder connection between the package and the PC board is very difficult to do. Vendors who support JTAG capability minimize the problems associated with PC board testing. JTAG can also be used to test internal logic. This is done by using one of the many internal modes that allow shift register chains to control and observe logic. All these testing techniques impact observation and control in the circuits only. Depending on the implementation they may add to or may subtract from test time. Do not confuse ease of test with length of test. Ease of test is the popular concern, but length of test is a major cause for cost overruns.

SELECTION OF PACKAGING

Vendor capability

Vendor capability for packaging varies dramatically, depending on the product and the process. Commercially supported packages include dual in-line packages, pin grid arrays, surface-mount packages such as PQFPs and PLCCs, small-outline integrated circuits, and many others. These come in both metric (millimeters) and English (inches) form. They may include plastic, ceramic packages, TAB, or strict die sales. Some packages come with items such as heat spreaders and heat sink capability.

Based on the complexity of the design and need for certain packaging, the customer may limit vendors based on their ability to package or handle the type of device that is needed for production. If the design must have high pin-count requirements, this may automatically eliminate certain vendors from the support list. In addition, check with the vendor as to how a particular package type will be tested.

There are many variations and allowable manufacturing techniques including room temperature testing, hot temperature testing, and automated and manual handling techniques. The equipment necessary to handle high-pin-count high-complexity packages in a manufacturing environment is expensive. Temperature testing capability is difficult to add if the vendor does not already have that capability. It is worthwhile to ask the vendor or potential vendors, if selecting the more complex packages, what the manufacturing flow would be for the device, and how they would guarantee performance over the entire temperature range. Chapter 9 has a detailed discussion of the characterization of devices and library elements. Characterization is the tool that can guarantee performance without actual temperature test.

Once again it is worthwhile to note the minimum and maximum die size requirements for these packages. Notice in Table 3-6 that minimum and maximum die sizes are required for every package. The minimum is based on the maximum allowable bond wire length, the pitches required for pads and bonding finger spacing within the package. The maximum is limited by a variety of factors including overall package size.

Notice from the previous section that a device of certain complexity on one process may not fit in a specific package. In addition to that, the design may wind up with too small a die for a particular package type on the more advanced processes. In the case of too small of a die, it is significantly easier to rectify as the vendor will simply enlarge the device to make the minimum allowable die size. These rules for assembly capability include bond wire lengths. The minimum allowable die size will vary from one vendor to another based on the reliability, and manufacturing rules and requirements of the vendor. Reviewing specification variations, such as different bond wire lengths allowable in the manufacturing process, will impact die size. Some vendors may allow waivers for bond wire length, but it is not a recommended change as reliability of the device can be dramatically decreased if the bond wire lengths become too long.

Chapter 11 of the book will discuss qualifications in detail, but here it should be noted that one of the failure modes of packages, especially plastic packages, is the physical movement of bond wires during the assembly process. Figure 3-7 shows the injection technique for a plastic package and the phenomena called bond wire sweep. This essentially forces the bond wires to move in the direction of the plastic flow. This may cause adjacent wires to short. Even if they do not short, it is quite likely that, if they come close, there may be a reliability problem over time. Do not change the maximum acceptable bond wire length to try to get more out of the device than what the vendor will support with its assembly process.

Table 3-6. Minimum and maximum die size by package.

AMKOR/ANAM ESTIMATIONS ON MIN/MAX DIE SIZES
PER PACKAGE TYPE

PACKAGE TYPE	LEAD COUNT	MINIMUM DIE SIZE (MILS)	MAXIMUM DIE SIZE (MILS)	COMMENTS
PDIP (300)	16	030 X 030	145 X 500	
	24	040 X 040	150 X 500	
PDIP (600)	24	040 X 040	390 X 600	Note 2
	40	070 X 070	390 X 650	Note 2
	48	100 X 100	390 X 650	Note 2
PLCC	44	070 X 070	400 X 400	
	68	105 X 105	700 X 700	Note 2
	84	140 X 140	900 X 900	Note 2
PQFP (Bumpered)	84	115 X 115	400 X 400	Note 1
	100	150 X 150	500 X 500	
	132	225 X 225	700 X 700	Note 2
	164	260 X 260	900 X 900	Note 1 & 2
	196	320 X 320	1000 X 1000	Note 1 & 2
QFP (EIAJ)	44	050 X 050	200 X 200	
	64	080 X 080	350 X 350	
	100	160 X 160	350 X 580	Note 2
	160	280 X 280	850 X 850	Note 2

Notes: 1. These packages are currently designed but not tooled at Amkor/Anam facilities.
2. Mechanical stress is a major concern on large die sizes. Amkor/Anam has no reliability data on die sizes larger than 500 mils.
Package estimates courtsey of Glen Koscal, Amkor Electroincs, and John Heacox, Amkor Electronics.

Alternative packages

Yet another capability that some manufacturers supply is the ability to package multiple die in a single package. This allows the interconnections from one die to another not to be limited by the number of pins on the integrated circuit. Die connections are limited by the complexity of the functions and the interconnect internal to the package. Routing signals on a ceramic substrate from one die to another is significantly simpler than routing signals outside of the integrated-circuit package.

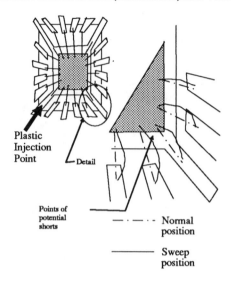

Plastic
Injection
Point   Detail

Points of
potential
shorts   — · — · · Normal
              position

              ————— Sweep
              position

Figure 3-7. Bond wire sweep.

When testing an integrated circuit with multiple die, it is even more important to add the testability hooks during the design process. This is to ensure that each of the integrated circuits, once manufactured in the module, are accessible, testable, and observable. The ability to rework one device in the package is a problem that most manufacturers do not attempt; there are reliability concerns when removing and adding die in the package. During some portions of the process, many companies allow no rework at all. For example, if a bond wire doesn't adhere to the device or needs to be removed, for considerations of reliability throw the device away rather than attempt to rework the device.

Packaging and pc board density

Density of the integrated-circuit devices on a printed circuit board is variable. Density is related to the number of pins in the package and the particular package type. Packages are available with 100-, 50-, and 25-mil pitches. Some packages have metric pitches of 1.0, 0.8 and 0.64 mm. These are available with single-, double-, and multiple-side packages or arrays and come in a variety of configurations. Typical packages are referenced as dual in-line, PLCC, quad flat packs, pin grid arrays, and many others.

Figure 3-8 shows the relative packaging density of pins per square inch of circuit density on a PC board. This is another tradeoff that is important to most system designers.

When selecting a package, the higher-density packages are more difficult to manufacture. The lead spacing is tighter and the handling problems are significantly greater in production for fine-pitch high pin-count packages. Although there may be some PC board savings, this may be wasted. If the manufacturer is not accustomed to the development, manufacturing, and production of high-pin-count small-lead-pitch packages, delivery may be a problem. There may be more rework and external costs than if a through-hole, mounted, low-density dual in-line package were used.

Chip-on-board programs have their own unique set of problems. This should be left to companies who understand the integrated-circuit assembly process. When doing chip on board, items such as die attach methods, encapsulation, bond wire types and inspection rules shift from the integrated-circuit vendor to the user. That shift brings the reliability and quality impact of poor process flow to the customer. Groups not familiar with integrated-circuit assembly should subcontract the assembly or proceed exceptionally cautiously.

MINIMUM AND MAXIMUM SPEED VARIATIONS

MOS transistors give a fairly strong performance based on temperature, voltage, and process variation. Most vendors have electrical test parameters used to screen wafers which are incorrectly processed and which are outside of the normal process distribution. Based on the width of the distribution of the acceptable process parameters, the variation of MOS processing speed is one factor that will impact the overall performance of the cell library.

In addition, voltage and temperature effects impact performance. A factor of as much as 5 to 8 delta may exist between the fastest product at cold temperature and the slowest product at high temperature over the simulation range. Vendors may present in their data catalog worse case numbers, or room temperature typical numbers, with derating factors or some variations of all of these. Figure 3-9 is a composite of multiple vendors, process variations, and the impact on the speed of the device. It shows three different manufacturers' processes and their derating factors for speed relative to slow process, hot temperature, and low voltage. Notice that the variation in speed can be substantial.

If you are looking at one vendor versus another and comparing a worst-case number versus a typical number, one may be headed for trouble. At the time of final simulation, when actually measuring worst-case performance, the designer will discover this problem. At this point you may need a major logic redesign to compensate for the speed that was inadvertently simulated incorrectly.

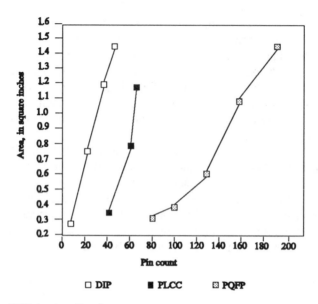

Figure 3-8. PCB density of packages.

VENDOR-COMPLETED DESIGN

Some vendors will actually take schematics, work with the engineering people, and complete the design. This total approach is far more expensive than what was cited in the table of typical costs for a design. However, it is an alternative for companies who want to do an ASIC design but lack the resources to do so. And it is also an alternative for those who do not want to learn the vendor's design system. In addition, there are design services companies who will act as an interface between the semiconductor vendor and the customer.

SUMMARY

The variations in the process, speed, performance, cost, density, and packaging capability from one vendor to another is dramatic. In addition some vendors support automated test capability and standards such as JTAG, built in self-testing, and scan, or LSSD capability. Direct access is not often considered a standard but is worthwhile. All these capabilities will be discussed in the next chapters in detail. They all will impact the ability to do a design, and their various advantages must be weighed in the decision process.

To help you decide, a simple matrix comparing parameters from one vendor versus another is useful. Table 3-7 shows a typical comparison of parameters for vendors and the weighting that might be chosen. The numbers here are for

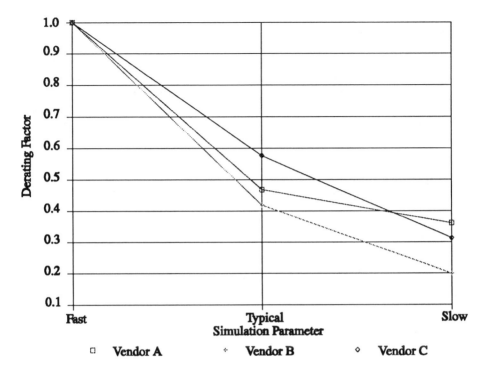

Figure 3-9. Speed and process variations.

illustration purposes only; each design needs to have its own weighting system and criteria.

Finally, remember that when a vendor is selected, the company is most likely going to stay with that vendor for quite some time. Learning the system for a design may take a considerable amount of time, which engineering and management may not want to repeat. The company will probably not be interested in the prospect of redoing all the evaluations and a new matrix for another design in the future.

NOT DOING AN ASIC

This basic understanding of test systems and ASIC technology allows the design engineer to review the decision to use an ASIC. Although the ASIC market is growing rapidly and ASIC designs are now considered the right approach, there are instances when implementation of an ASIC is not wise for the overall system.

Once again, evaluating volume, tradeoffs, and cost of implementation of the design may warrant the decision not to implement the ASIC device. Although an

Table 3-7. Selection matrix.

Factor	Maximum	Vendor A	Vendor B	Vendor C	Vendor D
TPT Protos	10				
Cost per unit	15				
Cores	7				
NRE cost	17				
Ability to manufacture	20				
Quality	15				
Test aids	20				
Reputation	10				
Experience	13				
Analog	5				
Package	15				
Military temp?	2				
TPT Production	8				
Training	11				
Total	698				

estimated 40,000 designs will be completed in 1990, there are several hundred thousand non-ASIC designs. Therefore, the total ASIC design population as a percentage of the total designs is a relatively small number. This means that a majority of the systems that are being designed are still being made with standard components.

4

TYING TOGETHER TESTING AND DESIGN

ASIC devices can be differentiated from standard products in three major ways: volume of production, time for development, and applications. Standard products are designed to sustain a high volume of production for a significant number of years. Figure 4-1 shows the relationship between standard product volume and time. As shown in this figure, many ASIC devices ramp to production and are out of production long before the standard product reaches high volume production. Standard products typically last several years and ramp to a significantly higher volume than ASIC devices.

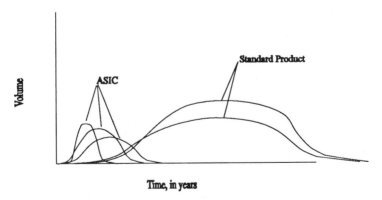

Figure 4-1. ASIC and standard product life cycles.

The test program for a standard product, may develop over many years. It is not uncommon to invest multiple labor-years in testing to ensure that the defect levels fall within an acceptable range. When developing a standard product, manufacturers often revise test programs repeatedly to include testing for manufacturing faults and application subtleties that have a positive impact on defect levels for the device. These changes or iterations in the test program may take several years to complete.

During this time of continuous revisions of the test program, modifications of the program for specifications changes and functional testing changes are done to ensure the highest possible quality. This particular form of continuous improvement is not usually available to the ASIC device user, because ASIC device volumes tend to ramp faster and product life cycles are usually significantly shorter than their standard product counterparts. The ability to revise the test program to lower defect levels is impeded by the rapid ramp of the production volume. The time necessary to locate problems, change the simulations, and adjust testing patterns to weed out the failures and then to transfer that pattern and program into production tend to slow changes. In many cases these changes will not or cannot be done for an ASIC device.

The applications-specific attribute of an ASIC device actually helps in test program and pattern generation. Test patterns and routines can be generated to check the device for use in its specific application only. The needs and concerns of applications unknown to the manufacturer at the time of design are not of the same importance in an ASIC device as they would be to a standard product.

The combination of these three factors—the application's specific design, the lower-than-normal volume, and the short life cycle of the product—changes the rules for test development of ASIC devices. The designer must take test capability and test design into account in the initial stages of the integrated-circuit design. The designer must assume that during this process there will not be changes to the design or testing of the integrated circuit to improve defect levels. If the design is implemented with testing in mind, very high quality levels can be obtained using present-day ASIC technology without test program changes.

DESIGN FOR TEST PROBLEMS

The basic problem in designing an integrated circuit for testing is to ensure that the internal nodes are accessible for controllability and observability. *Access for control* means that the individual states within the device can be set and reset based on external setup arranged by the tester. *Access for observability* means that once internal nodes are set or clocked—with the resultant logic computation—they are easily observable on the outside pins of the part. This ability to control and observe internal nodes will be the basis for all future fault grading, test programs, and test pattern generations for the device.

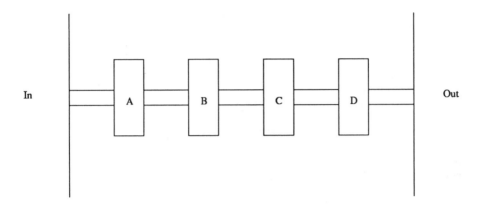

Figure 4-2. Traditional logic with no testability.

Nodes that are difficult to control or difficult to observe simply mean that the amount of test vectors and generation effort required to test that portion of the device is higher. Such testing efforts may be dramatically greater than in the highly observable, highly controllable states. It is also possible to design logic that is *not* testable.

An example of a device that is poorly observable and poorly controllable is shown in Figure 4-2. The four logic blocks shown in Figure 4-2 are block A, B, C, and D. A controls B, B controls C, and C controls D. D is directly observable by the outputs of the part, and A is directly controlled by the inputs of the part. This figure shows a serial interconnection technique for multiple blocks within an integrated circuit with no testability added.

Assuming that these blocks are made up of a combination of combinatorial and sequential logic, patterns would need to be generated that would exercise block A to control block B. Several patterns in block A would be required to set up block B simply to control block C. Then if this followed the normal algorithm from one vector per transistor of the device, block C which may have 2000 gates could require 10 to 100 vectors for each transistor tested.

If this same design was implemented with accessibility to all of the intermediate connections between blocks A, B, and C, patterns would be significantly easier to generate and shorter. This is shown in Figure 4-3. It is the same block as shown in the previous figure with multiplexors for control of input and output signals for each block.

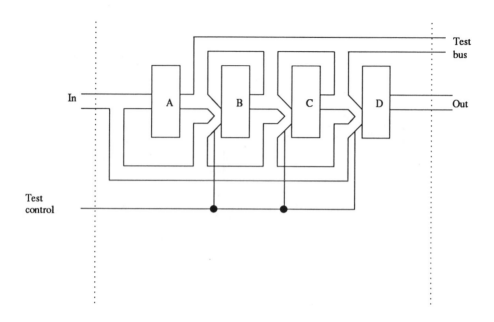

Figure 4-3. The same logic as in Figure 4-2 but with testability added.

The ability to generate the patterns for C instead of being 2000 times 10 or 100 data patterns, would be the simple case of 2000, thus dramatically cutting the amount of effort necessary to generate the vectors for the blocks. In a design of 20,000 or 50,000 devices, this overhead factor of 5, 10, or 20 to 1 can have a dramatic impact on test vector generation via control and observation of internal nodes.

This particular example solved the problem of observability and controllability by adding many extra pins to the device. Often that will not be practical. Therefore, one must resort to other techniques to control and observe the inputs and outputs of logic blocks within the integrated circuit.

TEST METHODOLOGY

When generating the test patterns for devices, it is important to note that testing the device is not the same as functional exercising of the device. Figure 4-4 shows a four-input NAND gate as well as the 16 possible combinations for functional verification and the 5 necessary combinations for functional testing. The same holds true for most logic elements. The number of states that the device must go through to ensure that each node is toggled, controlled, and

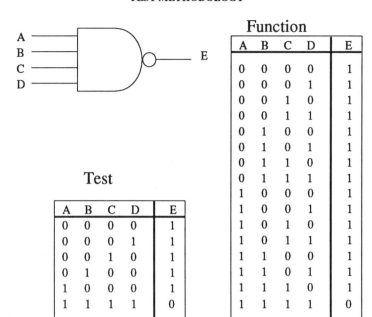

Figure 4-4. Test versus functional patterns.

observed is significantly lower than that to ensure that all possible combinations are checked.

This generation of patterns for test purposes should be limited to specific tests that exercise gates, blocks, and latches. This is to ensure their testability and not how well the device functions. Testing the performance of the logic in the test sequence will require different patterns for each individual path. Therefore, in the case of the four-input NAND gates, shown in the last figure, if the NAND gate is embedded in the circuit driving a latch, the paths that drive that NAND gate must be individually tested. Figure 4-5 shows the same logic gate with the appropriate drive circuitry prior to the gate and the four necessary paths to check out the functionality of the device. For the purposes of timing tests, only the worst-case path needs to be checked. The critical path is shown in the dark line.

Methods of testing an integrated circuit

For the engineer facing the task of testing of an integrated circuit, several alternatives are available. Table 4-1 shows the basic options available for testing an integrated circuit, varying from bench type of electronic test equipment to commercial ATE. The same table shows the relative relationship in cost for the

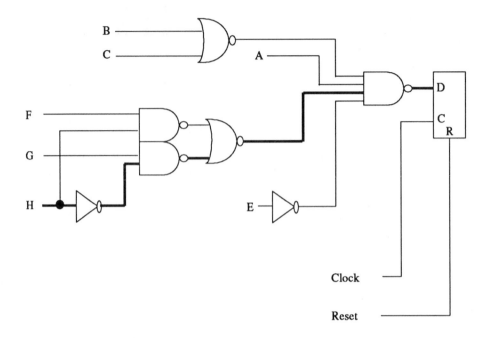

Figure 4-5. A four-input NAND gate in a circuit.

setup to test the device and the time necessary to test the device. The numbers in this table are approximations to illustrate the tradeoffs available.

Bench-type test equipment includes oscilloscopes, logic analyzers, and pattern generation equipment such as pulse generators, and logic pattern generators. This type of testing can be done in a stand alone socket with the integrated circuit exercised only by the electronic test equipment. To accomplish this the test equipment forces input conditions and measures output signals. It can also be done in the system. In this case the integrated circuit is working with the other circuits, switches, and displays that eventually make up the system. This fully exercises the device in its intended application.

Comparison testers are used to check the functionality of one device as opposed to another. They may also do a comparison of a model of the device generated in TTL versus the integrated circuit manufactured. This assumes that one of the two devices, the "golden unit," has been fully exercised in one of the other methods. Either the bench setup, system debug, or commercial automated test equipment could be used. Comparison testers are more thorough and faster than scope and logic analyzer work on a bench but usually do not have the accuracy of the commercial stored-response testers.

Table 4-1. Test cost tradeoffs.

	Bench	Verification	Production
Equipment cost	30K	200K	1000K
Time to develop a program	Months	Weeks	Weeks
Time to test one good device	Weeks	20-50 Seconds	1-2 Seconds

Stored-response testers are the final category and the most common category for testing of complex integrated circuits by most manufacturers. Stored-response testers are, again, broken into verification machines and production machines. The basic operating characteristic of a stored-response tester is that patterns generated in simulation are turned into the one and zero states in the pattern memory of the tester. The tester then executes patterns for the device under test response and compares the stored response with the device under test to determine its functionality. The major difference between verification systems and stored-response testers used in production is the ability to handle long patterns and execution time.

Fault grading

Fault grading is the metric of the test patterns to verify the effectiveness of the test pattern. The faults in a device can be monitored directly or by statistical averages. This is done to ensure that the test coverage is adequate. The simplest fault measure is a toggle count, which is the bare minimum that most manufacturers will accept. The toggle count simply observes every node in the device and ensures that it moves to a one and a zero during the exercising of the test patterns. This ensures only that the node is controllable. Any particular fault on the node is not necessarily propagated to the outputs of the device. Toggle count gives only a gross approximation of the testing. Without 100 percent toggle count, it is impossible to fully test the device. A 100 percent toggle count does not ensure that the patterns exercising the device cover all potential faults. If the patterns are generated with functionality in mind, toggle count can be a reasonable representation of the effectiveness of the test in some cases.

The next step up in complexity and quality of patterns is statistical fault grading. In this particular case a statistical evaluation process is used to determine whether the faults impressed upon the part are observable at the outside of the device. Algorithms are used to generate the faults, and based on the data patterns a probability is derived as to the probability of propagation to the outside pins of the part, where the faults may be observed. Statistical fault analysis is extremely fast in comparison to deterministic fault grading, but there is some uncertainty as to its coverage of the data pattern being adequate. Statistical fault grading cannot guarantee a 100 percent fault coverage pattern.

Deterministic fault grading is the most accurate for checking of the test patterns. It is also the most expensive. This method applies faults in the circuit either at inputs (the best way to fault the circuit) or outputs (*not* the best way to fault). In determinate fault grading, faults are forced on the circuit, patterns are run, and the output pattern is compared to see if the fault is observable. The test pattern used is the intended test pattern for the device. It should be noted that the time required for simulation of faults will also vary with the type of fault grading used. The cost is related to the method used and therefore can become expensive. One plan is to use progressively more stringent methods, so that time spent in the front of the process, when there are many faults, is inexpensive.

COST OF TESTABILITY

You must assume that the process of adding testability to the device and speeding up of logic will impact the size of the device and, therefore, its cost. In addition, the engineering effort to generate test patterns during the simulation process increases the complexity of the design and the amount of time necessary to execute the design. It is assumed that this compensates for the cost of generation and debugging of the test program after silicon has arrived. This should be weighed against the cost associated with errors in the test pattern that cause defective units to be shipped. The model that is often referenced on the cost of defects states that for every increased step in the process, the cost of repair goes up by an order of magnitude. Therefore, repairs and system problems that are caught in the field due to poor design and poor test procedures in manufacturing are the most expensive. They are substantially more costly than it would have been to throw the devices away at the test step during manufacture.

The important parameter in this equation is the estimated volume of the integrated circuit. For the case of very low volume, adding test logic on the front end may not be wise. Compare this with debugging the system with an oscilloscope and logic analyzers, and conventional system debugging tools on a bench may be a better choice. In this particular case, low volume is 50 units or so; in most semiconductor manufacturer's language, this is extremely low volume. In the case of extremely high volume (100,000 units per month), the cost of test circuitry added to the device over the lifetime of the product has a far bigger impact on total system cost. In this case, such as standard components that are manufactured for extremely high volume, it may be worthwhile to eliminate some of the testability circuits. Then supply additional vectors and debug time for the device during the introduction and manufacturing ramp cycle. Although this point is also subject to debate. Typical ASIC devices fall into the category of moderate volume (2000 to 50,000 per month). In these cases adding of debugging and repair time at the back end versus the time necessary to create the logic and additional cost per unit for the additional silicon area is probably a worthwhile tradeoff. Other variables sometimes enter the equation such as the type of market that is being served, the expected lifetime of the product, and the required defects per million for the end-system application after test. The three-dimensional relationship of volume, cost, and testability is shown in Figure 4-6. Each design must be analyzed to ensure that it is at the proper point for testability. This is checked relative to its expected volume and desired defect level. Choosing the optimum point of volume, cost, and testability is a difficult decision. Depending on the system needs, the different aspects may be in conflict with another. At times these factors can be orthogonal. There is no one answer that serves for all possible device types. Therefore, the increase in logic and design time must be less than the cost of a larger device times the volume plus the change in time to debug the device. In rudimentary terms, additional design cost must be less than production cost or:

$$\Delta \text{ Logic} + \Delta \text{ Time} < \Delta \text{ Cost} * \text{Volume} + \Delta \text{ Time}$$

On both sides of the equation listed above, there is an element of delay and a time-to-market aspect. Adding logic for testability increases the complexity and therefore the time necessary to generate the individual modules within the device. The same holds true for sample verification and production ramping capability. Without testability added to the device, debugging of the device and brute-force pattern generation adds to the production ramping time—an important factor to be reckoned with at the time of generation of the device.

Several studies have been done relating defects per million versus overhead of test logic, defects per million versus test time, and defects per million versus test development. All these aspects are interrelated. The decision process to determine the proper amount of testability for the particular device needs to be looked at in detail prior to the start of the design.

Cost tradeoffs

There is also a cost analysis based on the type of design to be done. If the volumes and costs in Figure 4-7 are representative of the vendors that are being used, an analysis could be generated based on volumes that show the effective cost per unit. This is a typical cost-volume curve. Initial introductions into production absorb a higher percentage of the cost of the device than the high-volume production. Figure 4-8 shows the relationship of cell-based, array-based, and full-custom-designed integrated circuits on a cost per unit versus the total production basis. It is the cost tradeoffs of full-custom gate arrays and standard cells on a dollar basis with volume, including the time for the design of the integrated circuit. The nonrecurring engineering (NRE) portion, as described earlier, usually represents only the charges for simulation, mask generation at the ASIC vendor, and generation of prototypes.

When deciding to convert from a gate array to a standard cell or full-custom implementation, one must be extremely certain of the volumes. The NRE charges for a semicustom device when spread over a relatively small volume of devices relates to an extremely high cost per unit. The cost of the engineer doing the design is included in the analysis for this figure.

Test effort tradeoffs

Implementation of test logic impacts many aspects of the device. Foremost is that the increased logic in the part directly increases the complexity of the design and the number of gates necessary to implement the design. As seen in previous sections, the logic that is containable within certain packages and die size combinations is finite. If the design is close to the boundaries, implementation of test logic may cause the design to exceed the available gate count or area allowed by the package process combination. In addition, implementing additional test logic during the design portion lengthens the design cycle somewhat by increasing the amount of logic and complexity of the design.

Increased complexity and lengthened design cycle require greater effort to generate the test program and may result in a decrease in the number of defective units shipped. Note again that the increased die size may slightly increase

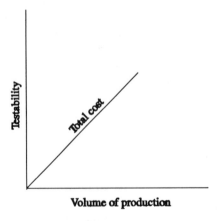

Volume of production

Figure 4-6. Test , volume, and cost relationship.

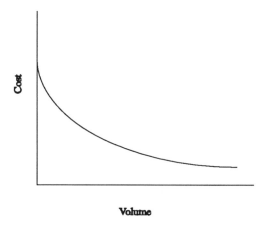

Figure 4-7. Typical cost-volume curve.

the cost of the device. It is assumed in this discussion that the increased cost of the device by adding testability will be paid for by the effort saved in generation of test programs. There is also an assumed reduction in the adverse cost of poor quality relative to system, customer quality, and reliability problems.

Once again the important variable in this equation is the volume. If the volume will be extremely high, one can afford to spend an extraordinary amount of time on the test capability of the device after the initial design process. The tradeoff of die size relative to complexity of the product and test capability is not anywhere near as important as volume. If the volume is relatively low or moderate, adding test capability to make blocks observable and controllable must be weighed against the engineering effort to generate the test program.

In both cases the implementation of test mode logic buys a tremendous amount relative to the quality and cost of manufacturing the system that the device will go into. For each step before a failure is found, you should use the model of a factor of 10 to calculate increase in cost. Obviously, delays are costly. Therefore, finding defects early in the manufacturing and test process, even though a larger die size is required to find them, may be extremely worthwhile.

Cost tradeoffs of ASIC devices

The preceding chapter contained a section that examined the costs of standard cells and gate arrays given different variables for NRE, volume, and production. Here it is also important to add that cost my vary greatly because design cycle of the devices vary. Depending on the expertise of the engineer doing the system design, there may be a significant amount of time invested in the design by the

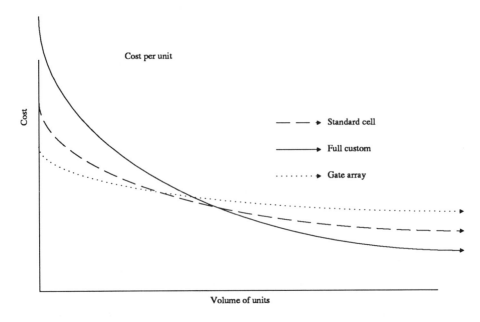

Figure 4-8. Cost of custom implementation.

user of the ASIC device. When this time expenditure is added to the design cost, one may have a very different picture. Assuming that the design cycle places the cost of an engineer at $50,000 a year, one can see the changes in the relative tradeoff between cell, gate array, and full-custom devices.

AN OVERVIEW OF TEST TECHNIQUES

The last few sections have reviewed the problems associated with test, and the lack of correct test techniques. Let us now focus on what to do to make testing easy. There are several methods available to use, and much printed information. Making the device testable will pay off in ease of implementation, lower costs, and fewer debugging hassles. The next few sections will focus on some of the test techniques used for ASIC test covering the cost of the methods and some of the pitfalls that may be encountered. In addition there are examples of commercial devices that use some of these test techniques.

Self-testing and test time

Several items can impact the overall cost of testing of an integrated circuit. They are total test time, the complexity of the vectors—therefore the amount of resources the machine for testing must have—and the length of the data patterns used for testing the device. Chapter 2 discussed the fact that all pins in parallel are always tested on an integrated circuit. Earlier in this chapter was a discussion on self-test techniques and how to exercise large blocks of logic. Note that self test techniques do not by themselves save any total test time or execution time in the test socket unless done in parallel.

One technique for effective self-testing is to allow the self-test portion of the circuitry to run while other portions of the device are being tested. Figure 4-9 shows the effect of self-testing done in the sequential manner and in a parallel manner. This figure represents serial and parallel starting of self-test techniques. The cumulative number of vector clock cycles as shown and the effect of parallel techniques dramatically reduces the number of clock cycles necessary to test the device. In this example, in the sequential manner, self-testing is implemented on block A. During this portion, a signature is done on the ROM contents and a signal is presented via test mode to the tester for determination of pass or fail logic. After this is completed, section B of the device is tested with self-test data patterns presented to the device in the serial manner.

The alternative presentation of the built-in self-test (BIST) is shown in the right portion of the figure where block A self-testing is started after the initialization and start sequence. Vectors for the random logic in block B are presented to the device. The end result of the self-test sequence is latched and stored for future reference, thus eliminating the problem of length of vectors equaling one another. The length of the vectors should exceed the self-test cycles; otherwise a tester decision of test in process must be made in the test program.

At the end of the test sequence for the random logic in block B, an additional vector is presented to the part to observe the output of section A, which is the self-test mode. This particular process allows the combining of self-testing and sequential testing in parallel. It also eliminates the need for additional formats and timing generators on the machines to perform parallel observation of all the blocks necessary for testing of the device. This is very valuable for a production test program for the debugging of silicon in the lab. It adds complexity to the debugging process, and it should be noted that it is not a requirement to execute in this particular manner. During debugging, each of the sections of the device are physically isolated from one another and there is no requirement to do the exercising of all blocks of logic simultaneously.

Scan and test time

When implementing a scan test in a circuit, the impact of the test must be understood. Scan uses serial vectors for the test of the device. This is a major change from direct access or self-testing in that the vectors are shifted in rather

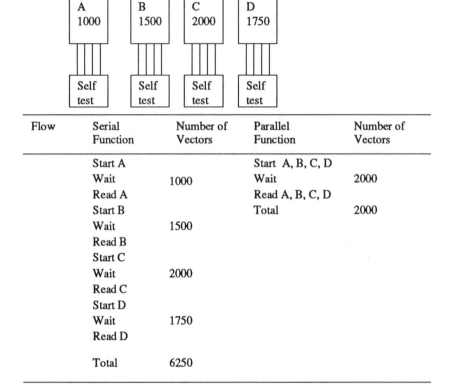

Flow	Serial Function	Number of Vectors	Parallel Function	Number of Vectors
	Start A		Start A, B, C, D	
	Wait	1000	Wait	2000
	Read A		Read A, B, C, D	
	Start B		Total	2000
	Wait	1500		
	Read B			
	Start C			
	Wait	2000		
	Read C			
	Start D			
	Wait	1750		
	Read D			
	Total	6250		

Figure 4-9. Self-testing and test time.

than presented in parallel to the part. On a small integrated circuit this is probably not a problem. Table 4-2 shows the impact on test time for scan versus parallel methods of testing. For this example, 10,000 gates and 40,000 vectors are to be tested. By the scan method, this takes 40 pins times 40,000 vectors at 1 microsecond (μs) each, or 160 ms. This pattern must be repeated for various combinations of V_{CC} and perhaps for input and output levels.

The same example using direct access takes only 40 ms. The impact to test time is considerable, and if the circuit takes significant vectors to test, the scan effect at testing can be quite costly.

Table 4-2. Scan and test time.

Parallel vectors		Serial shift pattern	
Vector number	Pin number 1111111111222222222233333333334 12345678901234567890123456789 0	Vector number	Bit number
1	0110101111101100011010100011101010001111	1	0
2	1100011010100011111111110110101111101111	2	1
3	1111111000101101011111011100011111100011	3	1
4	1111100110111111010100000110101111101111	4	0
5	0011010111110000011000100011010111110101	5	1
6	1101010000110101111101111101101011111001	6	0
7	1010100011111111110100001111111000111111	7	1
8	1010100011111010011111001111111000111111	8	1
		9	1
		10	1
		11	1
		12	0
		13	1
		14	1
		15	0
		16	0
		17	0
		18	1
		19	1
		20	0

TESTABILITY CHECKLIST

There are a minimum of three places in the design process where testability must be reviewed. The first portion is during the architectural overview of the device. This is the point when the basic block diagram is being partitioned into sections. What will be included inside the integrated circuit and outside the circuit is decided here. During this step it is important to look at large blocks of logic and how they will be tested, using the techniques that were described previously. These are techniques such as self-testing, scan-testing, direct access, or combinations of them.

The next point in the process that needs to have a detailed test review occurs after the initial logic design is completed or when the schematics are entered and the actual gate count is available. This is another review of testability and should be completed by again looking at the logic of the device for access control and observation.

The final point occurs during the final simulation and test pattern generation step. This is before the commitment to samples and is the last place where testability should be reviewed. Items to be evaluated in the testability review include those shown in the checklist as in Table 4-3. This checklist is primarily oriented toward testability and does not include good design techniques and good or bad design practices. This is just a basic testability checklist. Items such as the elimination of loops, asynchronous circuitry, feedback techniques, and nontestable logic are all design practices that are covered in detail in other material.

The important factor to note in the testability review and in the checklist is the controllability and observability of the blocks of logic. In reviewing the logic look for areas that have extremely complex sequences to set up or control. Does the circuit use long involved patterns to read or observe the logic? Are there areas that perhaps need more testability?

The step of checking and adding testability during the design of the project is the time in the process when you can easily add proper control and eliminate nontestable blocks. Once the device is committed to silicon, it is too late. If sections of the integrated circuit are not easily testable and do not function correctly or if they function differently than expected in simulations, considerable amounts of time and effort will probably go into analyzing them. Especially if they are not directly testable.

Table 4-3. Testability checklist.

Testability checklist	Checked
Accessible	
Observable	
No long test sequences	
What parts in parallel?	
Estimated margin to spec	
What paths are the most critical?	

DESIGN PRIORITIES AND TRADEOFFS

Many tradeoffs will be made during the design process. The priorities will be perhaps different, depending on the perspective of the individual looking at the device. The systems or ASIC design engineer, the management, and the individual responsible for testing the device may have a different perspectives on the priorities relative to the device. Items such as speed, performance, elegance of design, levels of integration, margin to specification, and pin count are often areas of concern.

Pin count, in particular, falls into a strange category in that it is a digital function. One either cares or does not care about pin count. The boundaries of packages and available pin counts are finite; this is especially true in plastic packages. When the design approaches the state where all pins on a plastic package are used, there is usually quite an incentive to stay within that package and not go to the next highest pin-count package. As seen in the previous chapter, minimum die size is dependent on the package type. If the ASIC being designed is a device targeted for a 100-pin plastic quad flat pack and the design uses only 94 or 95 pins, adding 4 additional pins to add test modes and test capabilities to a unit such as the P1149.1 is not a problem. Scan requirements for 4 pins is usually not a major concern. But if the design is already at 97 pins, implementing 4 more pins for P1149.1 can be a very painful decision. A 101-pin package is not commercially available, and this may require conversion to a 120- or 132-pin package. The cost of assembly of the package is directly proportionate to the leads, therefore a 132-pin package is about 1.3 times as expensive as a 100-pin package. And of course the die size may need to change also.

Understanding these concerns in the initial phases of the design and setting up priorities to ensure that the package testability and design all fit together in a cost-effective manner are important early decisions.

APPLICATIONS OF TEST TECHNIQUES IN COMMERCIALLY AVAILABLE STANDARD PRODUCTS

Thus far the discussion has focused on scan, direct access, and self-testing as primary techniques for gaining control and observability of internal nodes in an integrated circuit. All these techniques are used in commercial applications. In the next section we will evaluate two devices. The Motorola 68332 and the Intel UCS51 will be evaluated at length relative to the test logic and modes of operation of the devices.

The Motorola 68332[1] is a high-performance 32-bit microcontroller with reusable modules attached to a backplane like an internal bus. The device consists of a 1K * 16 RAM, a time processing unit, a central processing unit, serial ports, and various other functions. During the development of the test patterns of this device, several of the techniques discussed in the previous chapters were used.

Scan in the motorola 68332

Several areas of logic in the 68332 included scan testing capability. One section that was accessed by the scan path was testing of the programmable logic arrays (PLAs). The instruction decode PLA had 22 inputs and 27 outputs. The scan path ran through 19 of the inputs, the other 3 inputs were controlled by test mode bits in the test register. This required the PLA to be tested in 8 separate sequences. This arrangement was chosen to limit the input scan path to 19 bits since the serial pattern generator and the test module of the system integration module (SIM) is a 19-bit mini linear feedback shift register. A PLA can be tested deterministically or pseudoexhaustively with this scheme. During the testing of the PLAs, the microprogram counter acts as the output of the PLA instead of serving as an input to the ROM. Figure 4-10 references the block diagram of this section of circuitry; it shows the scan path in the 68332. Note the many elements included in the path.

Direct access in the 68332

The 68332 contains an intermodule bus that connects each of the macromodules together. This pipeline bus can handle up to 16 modules including 8 masters. It can be put into a mode which allows direct access to all modules on the intermodule bus. In this direct access mode, patterns for each block may be presented directly to the part and tested in a modular manner. These patterns become stand-alone patterns that can be used in any kind of mix and match, thus allowing any module or any combination of modules to be reused and connected in future devices of the same family. Because the modules are direct-access, fault grading patterns that are used for the generation of the modules are directly transferrable to any other device in the family.

---

1. *Portions of the following sections are reprinted with permission from the IEEE. Material is from a paper entitled "Testability features of the MC68332", by Harwood et al. Proceedings of the IEEE Test Conference, August 1989.*

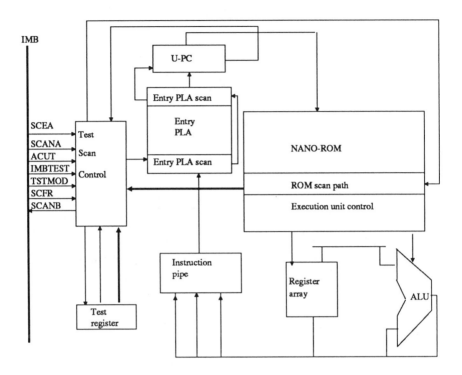

Reprinted with permission from Proceedings of the International Test Conference © 1989 IEEE.

Figure 4-10. The 68332 scan path.

Self-testing in the 68332

Built-in self-testing capability for the 68332 is applied primarily to ROMs and PLAs in the central processing unit (CPU) and the time processing unit (TPU). In this particular case, built-in self-testing is accomplished under the control of the CPU program. This sets up the test module for one test segment and then passes control to the test module during the test. Upon reset, the CPU program reads the reset status register and decides whether to continue testing. This self-test capability allows the CPU to verify the functioning of the larger arrays. Overall impact and additional test circuitry on the 68332 is 4 percent. This design allowed a highly modular, easily testable device that can be configured for many different applications.

Test techniques for the Intel UCS51 core

The Intel UCS51[2] Core uses a series of techniques similar to those described before. The UCS51 Core is a derivative of the Intel 80C51 microcontroller product. The UCS51 had a testability goal of ensuring that test development used for the 80C51 over its product life cycle would be applicable to the UCS51 Core. The UCS51 core is used for ASIC designs and can be implemented in a variety of configurations for many applications.

DIRECT ACCESS IN THE UCS51

For the UCS51, thirteen signals are required to interface to the outside of the device for direct-access testing of the core. With these signals, patterns for the core derived from the standard product can be presented to the core, and full functional testing in a direct-access scheme can be implemented with minimal overhead. This direct-access method is controlled by a test pin that will reconfigure this system for testing purposes. The activation of the test pin occurs only during the reset sequence and is designed not to accidentally occur in normal operation. Users do not have to generate all test sequences in their own data patterns. Figure 4-11 shows the pins and the test modes of the UCS51.

SELF-TESTING CAPABILITY IN THE UCS51

After the initial core is tested using the data patterns that were supplied by the standard product program, devices internal to the core and to its peripheral functions can be tested using self-testing techniques. Again, a test mode control signal is activated to allow the core to exercise the logic in the peripheral device in an isolated state and read back data for comparison purposes. For the core test, I/O ports are fed back similar to that described in Chapter 5. This allows the core to be used to test the actual I/O port functionality by exercising the I/O ports that are internal to the device.

---

2. *Portions of the following sections are reprinted with permission from the IEEE and Intel Corporation. The material is from a paper entitled, "Testability features of an ASIC Architectural Core," by Coleur and Cravatta, IEEE ASIC conference, September 1989. In addition material is taken from the Intel Introduction to Cell Based Design, January 1990.*

Test mode setup in the UCS51

The USC51 has a dual test register that controls the execution of the test modes of the device. The TEST signal is active during all UCS51 test modes. Pins may be mandatory or optional for test. The optional pins must be multiplexed to the outside of the device in test modes, thereby using the direct-access technique as described earlier. Actual selection of the test mode is done by decoding a register that controls access to the device. Modes of the test register decoding include ROM test, RAM test, internal I/O test, core test, address test and slave mode, and isolate the core from the device. There is also a mode that allows access to internal timing signals for the timing test in the device. These two examples show some use of the test modes to aid in testing. The final step is to review tradeoffs.

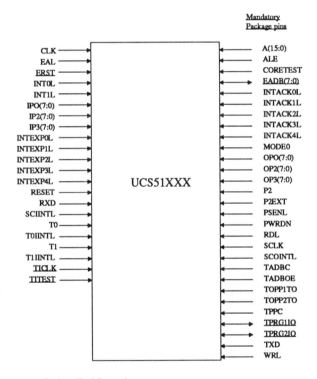

Courtesy of Intel Corporation.

Figure 4-11. UCS51 test pins.

OPTIMIZATION OF DESIGNS

Three factors are normally cited when optimization of design is considered: speed, power, and density or cost of the final product. There are arguments that these three considerations are in conflict at times, and in some cases it may be impossible to achieve all three goals for any particular ASIC design. Of course, the best design would offer infinite speed, no power, and no cost!

The previous chapters discussed testability at length. Testability should be added as a fourth parameter for optimization in the design process. It should be obvious by now, that if the design is not completed with testability in mind, speed, power, and density may make no difference to the final product. It may be impossible to tell if the device is functioning correctly and to gain the necessary confidence to release the product to production without proper testability. Testability, therefore, ranks with speed, power, and cost as one of the major concerns for a complete ASIC design.

METHODS OF PATTERN GENERATION

Now that the basic test techniques have been reviewed, there is also the aspect of generation of patterns for these test techniques. Although access to the nodes can be done via multiple different methods, some of these methods work better with different pattern generation techniques than other methods. For instance, direct-access test techniques work very well with functional test patterns that were generated for the block intended to be tested. The ability to generate long data patterns may be limited by tools such as assemblers or test program generation languages that allow rapid conversion of logic block understanding into patterns used for test.

Blocks that were implemented with a scan, LSSD, or modified P1149.1 approach to allow access into the internal nodes may lend themselves to exhaustive or pseudoexhaustive testing techniques. Most of these terms are explained in other textbooks and Edward McCluskey's book, *Logic Design Principles: With Emphasis on Testable Semicustom Circuits* is one of the better ones. Self-testing requires the implementation of the logic for testing in the hardware of the device. One of the problems with self-testing is that the patterns cannot be changed once implemented in silicon without a design change.

Some of the terms for testing may be confusing. Table 4-4 defines many of these terms as written by Edward McCluskey. These test definitions include many techniques for pattern generation that were not covered in this book. A more complete list is included in Appendix A.

After the decision is made to choose the method for access and observability (self-testing, scan, direct access, JTAG), a similar decision must be made on the type of pattern generation which will include random, pseudorandom, functional patterns, and hardware generated patterns. Note that the effort necessary to generate patterns for the testing of the device can be substantial, and if the rule of thumb of one vector for every transistor is followed, 50,000 to 100,000 vectors for a large ASIC design becomes a sizable task. In some cases, partitioning of vectors and testing responsibility may be divided up among multiple individuals.

Table 4-4. Test terms.

Built-in self-test, or BIST: capability of a product (chip, multichip assembly, or system) to carry out a functional test of itself.

Functional test (boolean test of a digital product): a sequence of valid input signals is applied to the device under test and the output response is compared to the correct response.

JTAG, or Joint Test Action Group: A collaborative organization composed of major semiconductor users in Europe and North America. Author of a proposed standard for a boundary scan register and test access port(TAP), which as evolved into the proposed IEEE standard P1149.1.

LFSR, or linear feedback shift register: a shift register with connection from some of the stages to the input of the first element though an exclusive-or gate

LSSD, or level-sensitive scan design: a scan-path technique for systems using latches as bistables and two or more independently controllable (two phase nonoverlapping) clocks.

Pseudorandom testing: testing in which pseudorandom binary numbers are used as test inputs.

Testing: the process of determining whether a piece of equipment is functioning correctly or is defective. Equipment can be defective because it doesn't function as designed or specified.

Reprinted with permission *IEEE Design and Test of Computers*, © August 1989, IEEE, p. 70.

SUMMARY

Tying all this together to ensure that it works in the test program for the functional testing of the part is vital. This challenge is the basic function of the ASIC designer. If not addressed at the time of the design, there is no way to correct the problem once the device has been manufactured. Inadequate attention to testing accounts in large part for the estimates of a success rate of only 20 to 30 percent in the industry today.

5

DESIGN FOR TESTABILITY

Before our discussion of the design process, let us examine a few of the test techniques that will assist the designer in the selection of logic and test circuitry to make the device testable. This chapter will cover the design aspects of self-testing, scan testing and direct-access testing. In addition, the impact of testing large blocks will be addressed.

SELF-TESTING

Self-testing is a capability that has become quite popular. In the previous chapter there was a discussion of adding logic for testing. Testing of RAMs, ROMs, and PLAs by mens of self-testing is relatively easy to do. This minimizes the problems associated with generating patterns for them. If, for instance, a RAM is buried deeply inside the part, and there is a reason not to add the overhead to bring the RAM to the outside for testing, there is an alternative: Build in self-testing capability.

Self-testing is essentially the implementation of logic in the circuitry to do testing without the use of the tester for pattern generation and comparison purposes. A tester is still needed to categorize failures and to separate good from bad units. In this case the test system supplies clocks to the device from the outputs of the device. The sequential elements are run with a known data pattern, and a signature is generated. The signature can be a simple go or no-go signal presented on one pin of the part, or the signal may be a polynomial generated during testing. This polynomial has some significance to actual states of the part. Figure 5-1 shows a typical self-testing technique implemented on a large block of random logic, inside an ASIC device. This block of logic is tested with self-test techniques. The number of unique inputs to the logic block should be limited to 20 bits so the total vectors are less than 1 million. You may need a lower limit based on the verification machine chosen.

Self-testing therefore is the ability of the part to exercise itself and to generate some kind of pass or fail signal. Self-test capability can be implemented on virtually any size block. It is important to understand the tradeoffs between the extra logic added to implement self-testing and the inherent logic of the block to be tested. For instance adding self-testing capability to a RAM requires adding counters and read, write, and multiplexor circuitry to allow access to the part. The access is needed not only by the self-test circuitry, but by the circuitry that would normally access the RAM.

Self-testing in a RAM

In the case of a small RAM of 1K bits, the added overhead is substantial relative to the transistor and gate count of the RAM. For a large RAM, such as 64K bits or greater, the ratio is not quite so great. This tradeoff of added gates for test

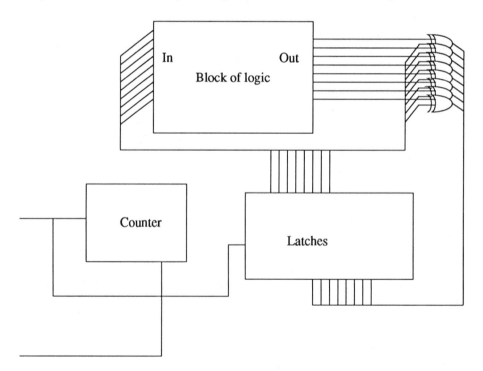

Figure 5-1. Typical self-test circuitry.

logic is shown in Figure 5-2; the added logic and an approximate gate count are also shown. This is a self-test technique for a 1K * 1 RAM. The sequence here would be to increment addresses, write data, and then read the data out to ensure that the proper data was stored in the RAM cell. This implementation writes a one or a zero to every location. It does not do pattern sensitivity testing. Because the test logic goes up linearly and the area of the RAM goes up by powers of 2, test logic overhead drops as the size of the RAM increases. Figure 5-3 graphically shows the added logic as a percentage of total logic for various-size RAMs. If access to this RAM is difficult from the outside of the device, you might be wise to implement self-testing capability. On the other hand, installing self-test capability for a block that is easily accessible would not be prudent.

In partitioning the circuit, keep in mind which blocks could easily be tested and how they could be implemented using self-testing techniques. Remember that the more difficult the block is to access from the outside world, the more it makes sense to try a self-testing technique. Self-testing can be very successful on many types of blocks. For instance, Figure 5-4 shows a ROM using self-testing techniques and the added circuitry. Notice the addition of a multiplexor for the address lines and an adder on the output to sum the individual bits. This ROM uses a basic implementation of self-test techniques. There are no row and column sensitivity checks in this ROM test. In this particular example, the address pattern for the ROM of zero through 1023 would be incrementally driven to the ROM. The outputs of the ROM are added to a base number, and at

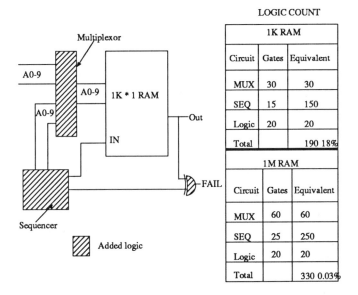

Figure 5-2. RAM self-test.

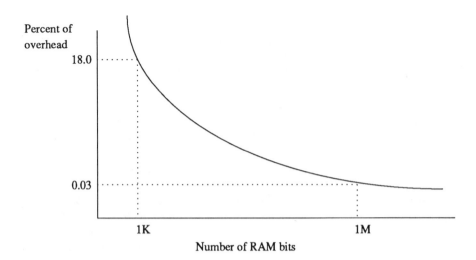

Figure 5-3. Self-test overhead.

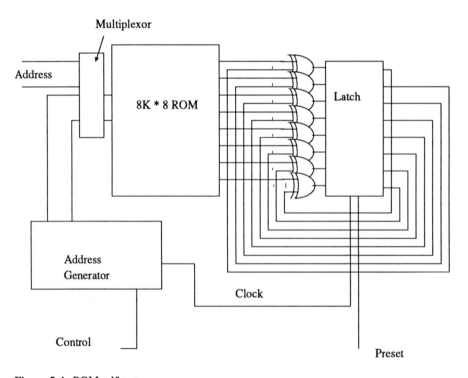

Figure 5-4. ROM self-test.

the end of the 1024 words of the ROM, the result should be zero. The initial calculation of the ROM start data in the summation register is, of course, dependent on the data pattern. If there is a change in the data pattern, the sets and resets of flip-flops for the ADD function must be changed. In this case the 1024 states of the part completely test this ROM regardless of how buried it is in the device. What is needed to make this a part of the normal functional testing of the device is to add control circuitry. This can be accomplished through a test mode to allow starting and stopping of counters. It will also enable the multiplexors and read capability to determine whether the register is at a zero state at the end of the test sequence. In this particular case the total amount of logic added is relatively small. On the other hand, some of the more complex blocks, such an arithmetic logic unit or multiply circuitry, may not lend themselves as easily to self-testing techniques.

SELF-TESTING CONCERNS

When implementing self-testing on blocks such as RAMs and ROMs, it is worthwhile to note the typical failure mode of semiconductor devices. Single-bit defects can easily be detected using self-testing techniques. Single-point defects in the manufacturing process can show up as a single transistor failure in a RAM or ROM, or they may be somewhat more complex. If a single-point defect happens to be in the decoder section or in a row or column within the RAM, a full section of the device may be nonfunctional.

The problem with this failure mode is that RAMs and ROMs are typically laid out in a square or rectangular array. They are usually decoded in powers of 2, such as a 32 * 64 or a 256 * 512 array. If the self-testing circuitry is an 8-bit-wide counter or linear feedback shift register, there may be problems. There are 256 possible combinations within the states of the counter, and this may be a multiple of the row or columns. Notice that there are 256 possible rows or multiples of 256 rows in the array, and 256 states in the counter make for a potential error-masking combination. If the right type of failure modes occur and a full row is nonfunctional, it may be masked. The implementation of the counter or shift register must be done with full-column or row failure modes in mind, or it can easily mask failures. This gives a false result that the device is passing the test. Simple techniques such as summing the device, the total bit contents, or toggling individual bits within the RAM or ROM may have more of a tendency to mask errors. The easiest way to prevent masking and such errors from passing test programs is to implement shifting or linear feedback shift register (LFSR) techniques. Figure 5-5 shows the implementation of an LFSR for a RAM. It shows a 2K * 8 RAM being tested with linear feedback shift register techniques. In this example the wrap-around techniques of the linear

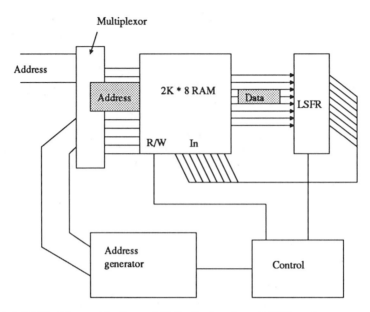

Figure 5-5. RAM self-test with a linear shift feedback register (LSFR) used.

feedback shift register prevents single-row and column failures from resulting in false pass conditions.

If the device that is being designed contains microprocessor or microcontroller capability, use the controller to exercise portions of the logic. Doing so allows test capability for the blocks that are embedded deeply inside. For instance, if a small microprocessor is embedded in the device with separate hardware multiplier accumulator circuitry, use the circuitry to run patterns on that microprocessor. These patterns would allow full testing of the logic without having direct access in the test program.

EXAMPLES OF SELF-TESTING

One use of self-testing capability is the ability to test I/O ports of internal microprocessors. Figure 5-6 shows a microprocessor buried internal to a large integrated circuit with I/O ports that access logic and are not connected to the outside world. One method of testing the device for ports 2 and 3 would be to add a test mode, which could make use of the serial shift register technique described later in the chapter. This technique allows shorting the pins of the I/O ports together. Shorting pin 0 of port 1 to pin 7 of port 2, pin 1 of port 1 to pin 6 of port 2, and so on through all eight combinations allows the connection of logic from the output of one output port to the input of the other input port. This

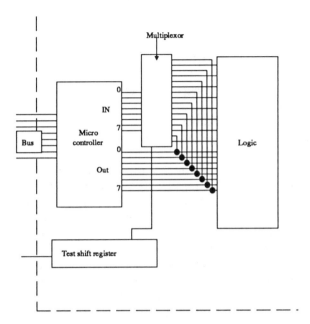

Figure 5-6. Microcontroller self-test.

method allows data to be directly latched onto one output port and read back on the other port. The input and output lines of this microcontroller are shorted together using a multiplexor. The input port can be connected to the logic or to the output lines of another port. Output line connections to input lines are based on the physical layout of the device to ensure that adjacent line shorts are not a problem. If the sequence of the pins is connected in reverse order (0-7, 1-6, 2-5, and so on), the possibility of accidental shorts showing false good data is minimized. This technique is used in production on commercial integrated circuits.

Another example of the same basic technique is to place multiple serial ports on a part. Shifting data out of one serial port into another is another use of self-testing. Transmitting and receiving the same data within the device allows for the testing of the baud rate generators, shift out, and receive circuitry. In addition, it checks latches and interrupts circuitry for the serial I/O port. This shorting of internal input and output nodes via the test mode is a very efficient way to gain access and observability to the logic of the device. It can be done without having to add a tremendous amount of extra overhead.

When using any self-testing technique, pay close attention to the failure modes of the device. Serial transmit and receive errors could be masked by other failures in the device. Care must be taken to ensure that a good functional test is made.

LSSD and scan design

LSSD and scan differ in implementation mainly by the set of rules used for design. LSSD has a strict rule for two-phase clocks that are non-overlapping, and logic that is not transition-dependent. For test considerations they may be treated in a similar manner.

Scan is a test technique that ties together all the latches in the part to form a path through the part to shift data. Figure 5-7 shows the implementation of a scan latch in a circuit. Note that the latch has a dual mode of operation. In the normal mode the latches act like normal flip-flops for data storage. In the scan mode the latches act like a shift register connecting one element to another. This is a basic implementation of a shift register or scan latch in a block of logic. Data can be shifted in via the shift pin or can be clocked in from the data pin. Clock line is the normal system clock and SCLK is the shift clock for scan operations.

Data for testing is shifted in on the serial data in pins of the device. Patterns for the exercise of the combinatorial logic could be generated by truth table or by random generation of patterns. These patterns are then clocked a single time to store the results of the combinatorial logic. The latches now contain the results of the combinatorial logic operations. Testing of the logic becomes quite easy, as the depth into the part is of no significance to the designer. Once the

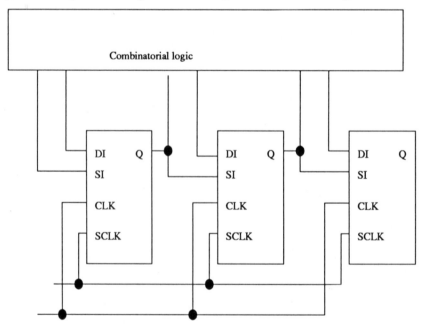

Figure 5-7. Use of a scan latch.

patterns are latched, the same serial technique is used to shift them out for comparison purposes. At the time of outward shift, new patterns are shifted in. Figure 5-8 shows this pattern shift for a circuit using scan. This is the actual implementation of scan in a small group of logic including the truth table associated with it and one state of testing.

Level-sensitive scan design (LSSD), or scan testing, can be used to exercise internal logic without having direct access to it. The latches form the scan chain. They also act as a storage element in their normal mode. The mode of operation is controlled by a test mode. Figure 5-9 shows a typical combinatorial logic design, set up in its normal mode of operation with scan capability added. It is a typical combinational block of logic with scan capability added around the perimeter of the logic, which allows full exercise of the logic. Typical testing of this type of logic would follow this flow. First, the device is placed in a test mode by using any one of a variety of methods, some of which are discussed below. One would be wise to use any of the methods that require specific pins to be set up on the part. A data pattern is shifted into the device through the data

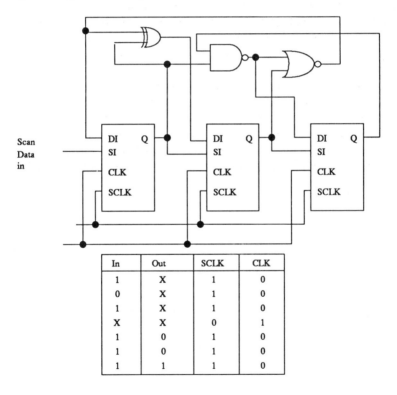

In	Out	SCLK	CLK
1	X	1	0
0	X	1	0
1	X	1	0
X	X	0	1
1	0	1	0
1	0	1	0
1	1	1	0

Figure 5-8. Implementation of scan in a circuit.

input to flip-flop A, then B, then C, and so forth. At the end of the shift sequence, all flip-flops are loaded with a data pattern, in this particular example, let us assume 10010. The device is allowed to clock one single cycle where the data from those latches propagates through the combinatorial logic and is relatched into the same latches. This saves the results of the internal logic calculation. The data pattern is then shifted out and compared to the expected data that was generated during simulation. At the coinciding time that the resultant data pattern is shifted outward, a new stimulus pattern could be shifted in for yet another combinatorial logic test. This process is repeated until all paths are tested. The procedure effectively checks all the data paths internal to the part without having to have general access. It also verifies that all the storage elements are working correctly and is a fairly effective tool for making test generation easy.

The tradeoffs for scan design orient around relative die size. Figure 5-10 shows the internal workings of a scan shift register latch versus a standard latch. This is the internal implementation of an LSSD latch and a D flip-flop. In normal operation, pins A and B (the scan clocks) are low, allowing the system data and system clock to propagate through the part. During the scan operation, the system clock is held low and clocks A and B are nonoverlapping, allowing data to be latched into the scan latch on clock A and latch out of the scan latch on clock B.

The size of the latch grows significantly in physical area. In this example the transistor count is 48 as opposed to 18 for a 160 percent increase. Edge-triggered latches have less of an increase in incremental area due to the initial complexity of the element. Assuming that there is a mix of combinatorial and sequential logic on the ASIC device, scan could increase total chip size by as much as 20 to 30 percent. What you acquire because of the 20 or 30 percent increase in die

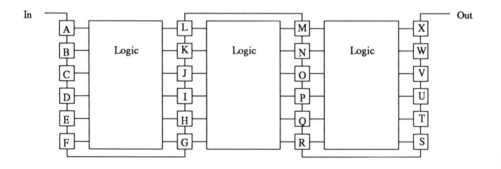

Figure 5-9. Typical combinatorial logic with scan latches added.

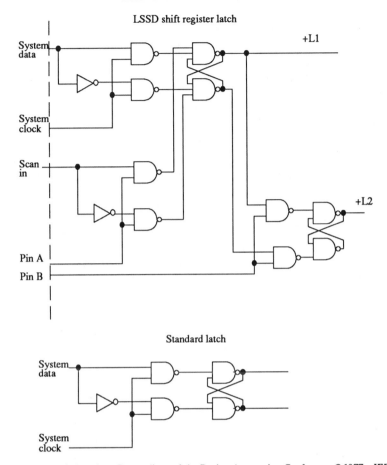

LSSD shift register latch

Standard latch

Reprinted, with permission, from Proceedings of the Design Automation Conference ©1977   IEEE.

Figure 5-10. LSSD shift register latch and standard latch.

area is a highly testable part that is easily accessible for all internal nodes. While this does not provide the smallest possible die size, it is a very testable die.

Another advantage to scan design is that automated test vector generation is readily available because the combinatorial logic is accessed so easily. This may be simply done by random patterns being implemented and switched into the shift register, and faults monitored, until the metric has met some predetermined criteria for fault coverage. This does not check the functionality of the part relative to the system specification, but it does ensure that any logic that is designed is correctly tested to ensure catching failures.

DESIGN TECHNIQUES FOR SCAN

If multiple blocks are implemented within the integrated circuit and each contains its own scan path, there are several alternatives for gaining access to the individual paths. One method is to daisy-chain the blocks together as shown in Figure 5-11. This illustration shows a serial scan chain implemented in blocks of logic, which are arranged in sequential fashion in this particular drawing. In this case one pin is used for scan in, and one for scan out. Test modes and clock pins will still be needed. This example could result in very long chains with long vector sequences to shift the data in and out. Long vector sequences mean long test time, and as wold be expected may impact test cost. The advantages of using the long vector chain is that it consumes only a few additional pins on the integrated circuit to gain access to the scan path. If pins are available, an alternate technique would be to wire the multiple scan chains to the output of the device in parallel. Figure 5-12 shows the same integrated circuit with the individual paths brought out to separate pins on the device. In this particular case three logic blocks with three scan chains are brought to the outside of the package. This setup consumes a total of 8 pins for the shift-in/shift-out information on all scan paths.

If pins in the integrated circuit are precious, one alternative is to mix the direct-access and parallel scan techniques together on the same device. This particular technique, shown in Figure 5-13, would allow the test mode to select multiplexors on the output pins, thereby selecting scan paths where data could

Serial Scan Chains

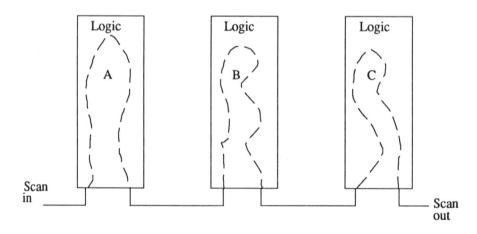

Figure 5-11. Serial scan logic.

Serial Scan Chains

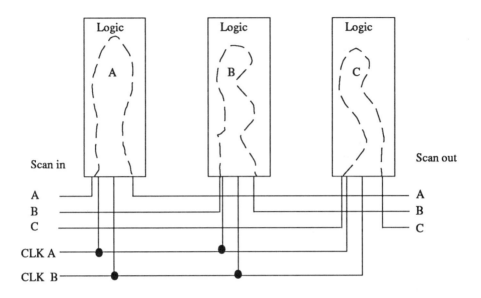

Figure 5-12. Parallel scan paths.

be shifted out. This technique is a combination of parallel scan-in capability and a direct-access block controlled by a test mode. In one operation the direct-access mode would be allowed access to the output pins of the part; in the other mode of operation, the scan chains are accessed directly. If there were sufficient pins, all the outputs could be brought out and tested in parallel if they did not interfere with one another.

The advantage of parallel testing for scan paths is that it allows vector sequences in scan chains to be run in parallel. In this case the simulations must be done in the right manner to accommodate possible difference in scan lengths. This is yet another design consideration that may impact the ability to test the device. Another potential for scan, if the patterns were random in nature, is to use the same stimulus vector set for all scan chains. This would mean applying the same pattern in parallel to all chains. Although it would require separate outputs for verification, it could be done. In this case all the scan in lines are shorted together to one input pin.

Serial Scan Chains

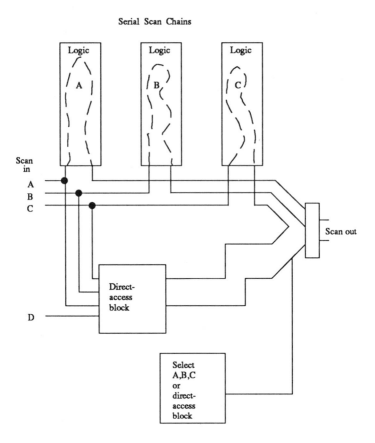

Figure 5-13. Combined scan and direct-access techniques.

Scan benefit

One of the side benefits of using scan test capability for integrated-circuit testing schemes is that it paves the way for automatic test pattern generation. There are several different theories available and implemented in practice related to exhaustive and pseudoexhaustive testing techniques that allow the generation of algorithms to check complete functionality of the logic. This ensures that the test sequence exercises the device fully under test. Not all vendors support this capability.

SCAN TEST IMPLEMENTATION

After the decision is made to implement scan testing on a portion of the logic within the integrated circuit, one must analyze the impact on the device test. Examine the scan chain and the test sequence that will be used for the device to ensure test time is not more than is reasonable. Scan chains data usually consume a lot of vector storage space in the test system. Some machines with specific memories or modifications to implement scan testing do not require the use of an extraordinary amount of vector memory. These machines typically have a separate memory that is configured for serial shifting applications. Figure 5-14 shows the configuration of a test system as referenced in Chapter 1. This

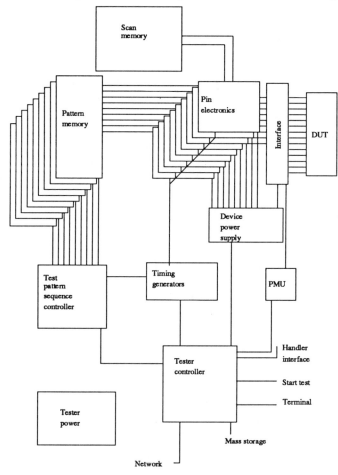

Figure 5-14. Scan memory in a tester.

system has a modification of an additional memory for the storage of scan data patterns. The addition of the scan memory at the top of the diagram allows long vector streams to be impressed on a few pins. It is important to note when wiring interfaces that the scan memories do not usually go to every pin of the pin electronics section. In the normal configuration this memory is long in vector depth and narrow in width. The memory can be accessed by only selected pins of the device under test using only certain channels of the tester. This memory then allows long strings to be shifted without consumption of the main vector memory for functional testing. Note that Pin 1 accesses the first data bit. There are fixed assignments for scan data to certain DUT pins. This must be remembered when wiring sockets.

Sample shifting techniques

Yet another method of observing internal nodes is by introducing a sample shifting technique that allows sampling of internal nodes to be controlled by an external clock. This setup, in essence, captures the nodes during a portion of the vector sequence. This capability does not require the scan-in/scan-out capability as in normal scan logic; it requires a shift-out capability only. Functional patterns are generated that exercise the device, and internal nodes are captured to ensure the functionality of those nodes at the time of execution of data patterns. Figure 5-15 shows a scan-out-only data pattern circuit added to a logic diagram.

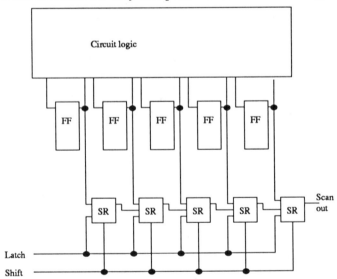

Figure 5-15. Scan-out logic.

The normal flip-flops are controlled by the logic within the device. The scan latches on the lower portion of the diagram allow latching of data and shifting of data. Although this is a good technique for observation of internal nodes, it allows no control.

Comparing this technique with Figure 5-16, which was the scan test capability, you will notice that fewer pins are consumed for test modes and less logic is added. In addition, no control functions have been provided that would normally be derived from scan testing. Figure 5-16 is the same block of logic implemented with scan latches. This is a more thorough functional test which allows control and not just observation.

DIRECT ACCESS

Direct access is a method whereby one gains access to the device logic block by bringing signals from the block to the outside world via multiplexors. Data is then forced into the block directly and the outputs are directly measured. This is one of the simplest methods to check devices for logic functionality. This particular method would supplement previous testing methods by allowing the user to impose data patterns directly on large blocks of logic. The same feature holds true for output observation. As shown in previous examples, setting up test

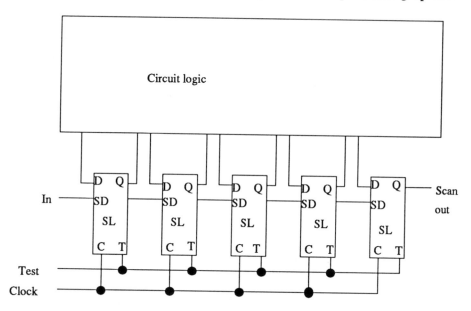

Figure 5-16. Logic with scan latches.

modes to exercise a part is quite easy to do if planned ahead of time. In the direct-access scheme, access of a test mode would force certain logic blocks via multiplexors to have access to the outside pins of the part. One could then drive the data patterns to the input pins, compare output pins of the part, and measure the access, status, and logical functionality of the device directly.

Figure 5-17 shows implementation of direct-access test techniques on a block of logic. During normal operation the block of logic has inputs driven by the logic block A, and outputs are connected to the logic block B. In the test mode, the two multiplexors are switched so that the input and output pins of the device can control and observe the block of logic directly.

IEEE P1149.1 or JTAG

When implemented in an ASIC device, JTAG, or IEEE 1149.1,[1] allows rapid and accurate measurement of the direct connections from one device to another

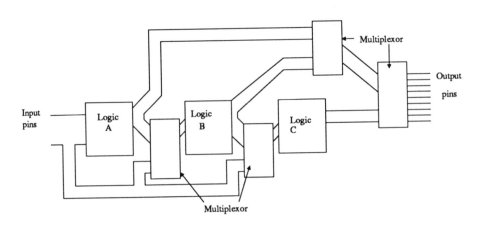

Figure 5-17. Direct-access implementation.

---

1. *Portions of the following sections are taken from the IEEE Standard for a Testability Bus, ©1989, IEEE.*

on a PC board. It is a very simple task to add if the library contains the basic elements. Many cell libraries now contain the necessary cells to allow easy implementation of JTAG testing. Figure 5-18 shows the JTAG technique in a small device, thus allowing accurate measurement and detection of solder connections and bridging on a PC board. This technique allows the shifting of data through the input and output pins of the part to ensure correct connections on the PC board. The four required pins are shown at the bottom of the drawing, and all the I/O ports are modified to include JTAG latches. The internal logic of the part does not need to change in order to implement JTAG capability.

JTAG test capability, or IEEE P1149.1 standard test access port and boundary scan capability, allows the designer to implement features that enhance testing. Both PC board test capability and the internal logic verification of the device can be facilitated. For systems where remaining components on the boards are implemented with the JTAG scan approach, adding JTAG to an ASIC device is a wise step to take. Some manufacturers have the building blocks required to insert JTAG easily into the integrated circuit.

The most important tradeoff that the system designer must face is the addition of the incremental four pins necessary for test capability. If the four pins are added for JTAG capability, this scan architecture allows control of most other test modes that were described. Direct-access and LSSD techniques can be controlled directly by the test access port of the JTAG specification.

The four independent pins mentioned above that need to be added to the integrated circuit include:

1. TMS, test mode select pin—Most often a signal common to all integrated circuits on the board, which controls the test logic operation.

2. TCK, test clock pin—Allows the shifting of data in on the TMS and TDI pins. It is a positive edge-triggered clock with the TMS and TCK pins that define internal state of the device. A sixteen-state finite-state machine, with certain states allocated for functions within the JTAG specifications, is controlled by the clocks and test mode. Other states definable by the user for identification, or testing, or for controlling other test technique purposes are also available.

3. TDI, test data in pin—The serial data input for the shift register. It connects all the storage elements around the outer edge of the integrated circuit. This allows shifting of one and zero data patterns to control individual pins for input function, output function, and the associated levels. It also allows the scanning of data through the device to the next device on the PC board network that the JTAG network is built in.

4. TDO, test data out pin—The serial data output. Through it, data is shifted out of the device on the negative edge of the TCK pin. In normal operation the data out would be the stored data from the latches that were placed on the inputs and outputs of the

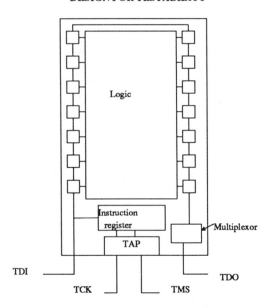

Figure 5-18. Basic P1149.2 implementation.

device. In normal implementation on a printed circuit board, TDO of one package is connected to TDI of the next package.

Figure 5-19 shows a PC board with three devices on it and the appropriate wiring for the 4 pins to allow test mode access using JTAG capability. Note that in the drawing, TMS and TCK are common signals routed to all devices. TDI and TDO are serial pins that start on the boundary of the board, shift serially through the three devices, and exit at the boundary of the board. Functions that can be checked with JTAG capability include pin-to-pin connections on PC boards, pin-to-pin shorts in the printed circuit boards, and internal logic of the integrated circuit. In addition there is a scan feature that allows monitoring the internal states of the part while in normal operation.

The external mode allows the scan latches to look outward, primarily at the device interconnections on the PC board. If there are any open PC board traces or solder bridges, they are detected using this capability. Figure 5-20 shows the same circuitry with an error in the PC board. The two pins are shorted together and one is open. As data is shifted in and strobed, data transfer from one device to another from the output of one set of pins to the input of another set of pins would be an error and when the data is shifted out, the opens and the shorts would be detected. Figure 5-20 shows the data pattern shifted in and some of the pins connecting the two devices. Pins 8 and 9 are shorted together while Pin 10 is open. The test sequence would then drive data into all the devices and all the

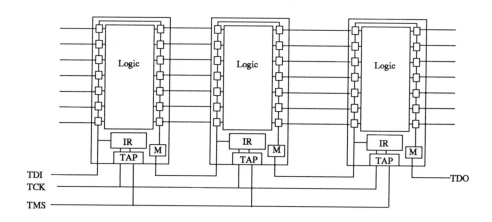

Figure 5-19. JTAG use on a PC board.

latches in parallel. A clock is supplied that allows input on the second device to receive the data and store it in the shift register. The data patterns are then shifted out. During the shift-out sequence, shown in Table 5-1, the expected data for a proper operating system is shown. Also shown is the data for the device with the short and open. Note the transition failures, which show that the in-

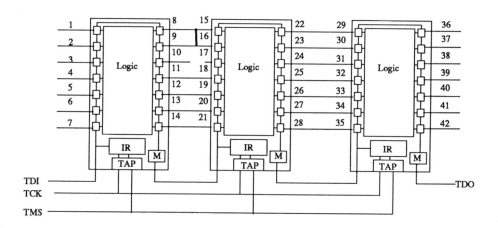

Figure 5-20. PC board error detection using JTAG.

dividual pins and the type of failure mode are easily identified. The internal test mode allows the user to shift a data pattern and clock patterns into the logic of the device. If the device is a combination of combinatorial and sequential logic, it may require multiple clocks. If there is built-in self-test capability, it may require many clocks to exercise the logic. At the end of the test sequence, data from the logic is stored in the output latches and shifted out. The results of the data pattern are then monitored by the tester and can be used for a pass or fail determination of the logic.

Figure 5-21 shows a block of logic using the scan-in/scan-out approach and two specific functional tests taking place. The first is combinatorial logic con-

Table 5-1. JTAG shift patterns.

Shift-In Pattern		Shift-Out Pattern		
Pin	Data	Pin	Expected	Actual
1	0	1	IN	
2	1	2	IN	
3	0	3	IN	
4	1	4	IN	
5	0	5	IN	
6	1	6	IN	
7	0	7	IN	
8	1	8	X	
9	0	9	X	
10	1	10	X	
11	0	11	X	
12	1	12	X	
13	0	13	X	
14	1	14	X	
15	0	15	1	0 ←
16	1	16	0	0
17	0	17	1	0 ←
18	1	18	0	0
19	0	19	1	1
20	1	20	0	0
21	0	21	1	1
22	1	22	X	
23	0	23	X	
24	1	24	X	
25	0	25	X	
26	1	26	X	
27	0	27	X	
28	1	28	X	
29	0	29	1	
30	1	30	0	
31	0	31	1	
32	1	32	0	

necting Pins 15, 16, and 20. The second is the built-in self-test capability for the
buried RAM block using Pins 17, 18, and 19. Figure 5-22 shows this in detail.
This figure shows the implementation of built-in self-test internal logic shift
testing and JTAG external testing in a single device. The built-in self-test for the
RAM is initiated and monitored using the control and shift pins. The combinatorial logic in the center of the drawing is directly tested by an internal
JTAG scan ring. In the case of the built-in self-test capability for the RAM, this
testing was implemented as described above, including necessary counters, pattern generation, and fault-detection capability. The output of the circuitry which
determines or monitors the accuracy or faults of the RAM is presented on one
pin of the shift register in this test mode and can be shifted out. During the shift
sequence, a data pattern is shifted in, as shown in Figure 5-22. The proper
operation of the logic is observable during the shift-out sequence, as shown in
the same figure.

The details of the implementation of the RAM self-test and the logic test are
also shown in Figure 5-22. The shift register ring controls the start of the test,
can access test in process, and reads the fail signal. Pins 15, 16, and 20 are used
to test the NAND gate as part of the logic function. Errors in the combinatorial
logic show up as bits in error in the data string. This is denoted by the A in the

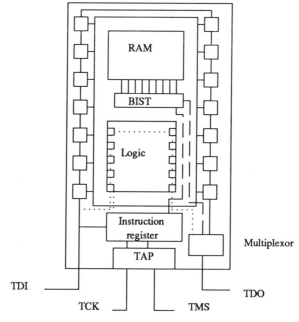

Figure 5-21. Self-testing with JTAG.

data path and the bit signifying pass and fail of the self-testing capability appears on bit B. For this sequence to work, the data pattern must be shifted in and the number of clocks necessary for the self-test capabilities for the RAM need to be fully presented to the device to ensure that it is fully exercised. The normal mode of operation of the boundary scan disables the latches, and all the devices do not impact the operation of the device. The device acts as it was originally designed and the four pins for the JTAG capability may be inactive.

The final mode of the JTAG capability that will be discussed is the sample test mode. This is effectively the scan-out capability, as presented in the earlier sections. It allows the latching of data at a selected point during the execution of the test sequence or functioning of the device. That data pattern can then be shifted out on the data-out pin of the JTAG port. This is a good capability for

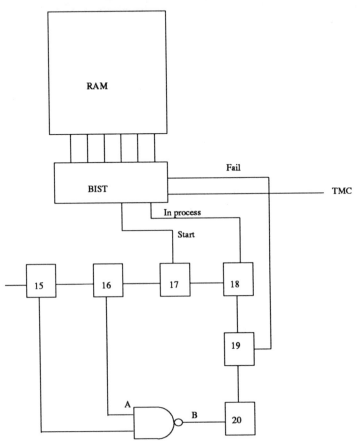

Figure 5-22. Details of self-testing and scan logic.

monitoring functionality and for determining the proper internal functionality of states during debugging. During normal test of the integrated circuit, it is probably not used.

The internal and external test modes of JTAG capability are usually done under slow speed: 1 MHz is typical. The sample test mode does not have the limitation of running at a slow frequency. Data can be strobed at the operating frequency of the device and shifted out.

Mixing of IEEE P1149.1 and other test techniques

It should be obvious that the JTAG capability allows for multiple user defined inputs for shifting of data patterns in and out of the device. The discussion of direct-access cited that a required number of pins be dedicated for test mode. Then multiplexors can be switched on the inputs and outputs for directly accessing and testing large blocks within the integrated circuit. This control of the test mode could be done using the JTAG capability. The blocks could be directly controllable and observable through the other input/output pins; it is not necessary that the pins be shifted in by JTAG and out by JTAG scan chains.

The same holds true for built-in self-testing techniques, as described in Figure 5-21. The initiation and monitoring of the built-in self-test capability could be done via the JTAG register. It can also be done by individual pins or by selecting a test mode that allows access through external pins to the start, stop, and monitor pins of the internal test mode.

METHODS FOR IMPLEMENTATION OF TEST MODES

Test modes allow the design engineer the access to individual nodes in the device by routing signals that are normally used for one function to another function for control and observability. There are several different techniques for implementing test modes in a device for a production test program—all of which have been tried and proved in production, some better than others for special cases. The idea to use test modes to take essentially one or two pins to control the functionality of internal logic that controls the testability of the block.

In the case of multiple-use pins, one pin is used to switch the functionality of input and output devices for the access of internal nodes. This includes the routing of input signals to the additional input pins. The setup used in Figure 5-23 with logic blocks A, B and C is implemented with a simple test mode control circuitry on these blocks of logic. This is the implementation of a single pin to control test modes for a block. Block B can be directly accessed by the input pins and directly observed on the output pins by control of a test pin. That test pin switches multiplexors on the input and output of the device of the block.

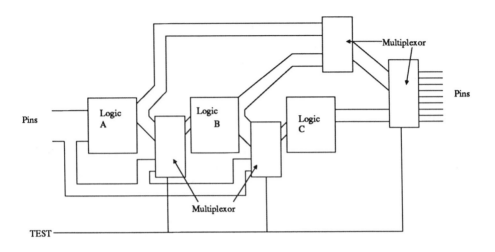

Figure 5-23. Single-pin sets mode control.

Notice that Pin 1 is used to control the test mode. In this case inputs to block B are routed around block A by a multiplexor, and the outputs from block B are routed directly to multiplexors on the output pins for direct observation. This allows direct access of the pins of the individual block via the pins on the device under test. This particular case uses a technique called direct access. Similar use of a single pin for test control would allow the implementation of self-testing, block observation, and other methods.

Another method for implementation of test techniques is to set up a shift register to control a test key word within the device. In this particular example, again as seen in Figure 5-24, the register of seven bits is implemented. This is the same drawing as in Figure 5-23, but instead of a single pin, there is now a shift register controlled by an input signal and a clock. Data is shifted into the shift register, and Bit 4 is used to control the test mode for block B. A clock and data line are available to the outside of the device. The data line could actually be multiplexed with other pins on the part. The clock line is a strict test mode pin dedicated only for testing. The clock is then cycled and the data pattern is shifted in to set up a particular test mode. In this case all ones represent the observation and control of block B, and block B inputs are routed around block A directly to pins of the device. Block B's outputs are routed around through multiplexors to the outputs of the device. The normal reset function of the device forces the

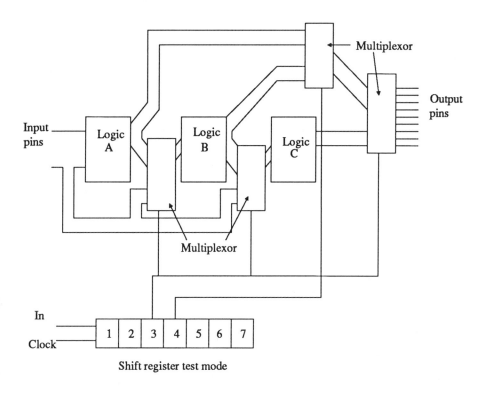

Figure 5-24. Shift register test mode control.

device out of the test mode to ensure that there is no spurious operation of the test mode during normal operation of the part.

One of the advantages of the shift register technique is that it allows many pins to be observed with direct-access schemes and yet use only one or two pins for input data to control the part. In complex designs where perhaps 10, 20, or 30 logic blocks implement the full integrated circuit, it is quite easy to access and observe all of them using this test mode without having to add an unusual amount of excess logic. A continuation of test pins and a shift register pattern allows the same shift register and control function to be used for items such as self-testing setup, scan rings, and any other control or observation function that would enhance the testability of the device. This use of additional logic to ensure that the device is functional will be emphasized over and over throughout this text. The design engineer has the responsibility to ensure that the test modes are implemented in a correct fashion so that the logic can be controlled and observed directly. This decreases the impact of defects, poor testing, and quality problems that a customer may inherit.

Gate complexity versus test time

Using parallel versus serial testing techniques has a big impact on test time. Assuming that the device has a total of 20,000 gates would give somewhere between 20,000 and 80,000 necessary test vectors. If the test vectors were a 40-pin-wide device, the total number of bits generated for testing of the device would be 3.2 million in the worst-case situation. If this is tested with LSSD or scanning techniques that are done in the serial fashion, 3.2 million bits shifted in at a 1 MHz rate means a 3.2 s test time. If you combine that figure with multiple values of V_{CC} and perhaps margin testing, test time could take several seconds in this particular case.

Using the techniques of parallel testing, if the shift registers could be fed in parallel or if some of the logic were implemented with direct access, the same 80,000 vectors implemented for two different values of V_{CC} would be 160 ms of test time. This tradeoff is significant, and most manufacturers have rules relative to the length of time that a device is allowed to execute in a test socket. As stated before, the machines that are used for production may be very expensive, and test time can represent a major portion of the integrated-circuit cost.

Tradeoffs

Notice that all these techniques add either additional logic or additional time delay to the circuit. Direct access puts a mutiplexor in the path of any data, thus slowing the circuit down by a few nanoseconds. LSSD latches are more complex than regular latches and therefore take up more silicon space. JTAG adds the latches to the output pads and consumes silicon area. It must be decided to add the circuitry, extra cost, and time needed to do the testing. Bear in mind that the logic is being traded off for easier access to the part for some added observability in the part.

Our attention will now focus on the tradeoffs of libraries, packages, and blocks for the implementation of the design.

LIBRARY SELECTION

The selection of libraries is perhaps one of the biggest aspects of ASIC design. The cost for the libraries are inversely proportional to the geometry size. A small feature library implementing a large design is significantly more expensive on a wafer basis than a larger geometry library, but there is a gain in performance and a gain in device density. One of the tradeoffs that must be made is which library to select.

Whenever you develop an ASIC design, the situation may arise where it is significantly cheaper to add logic to make logic decisions in a parallel manner rather than go to a higher-speed, higher-density process. This is especially true for the case of bond-pad-limited designs. Bond-pad-limited means that the bond pads are the factor that determines the minimum size of the silicon device. Referring to the die size Table 3-3, shown here in Table 5-2, you can see cases

Table 5-2. Minimum and maximum die size.

AMKOR/ANAM ESTIMATIONS ON MIN/MAX DIE SIZES PER PACKAGE TYPE

PACKAGE TYPE	LEAD COUNT	MINIMUM DIE SIZE (MILS)	MAXIMUM DIE SIZE (MILS)	COMMENTS
PDIP (300)	16	030 X 030	145 X 500	
	24	040 X 040	150 X 500	
PDIP (600)	24	040 X 040	390 X 600	Note 2
	40	070 X 070	390 X 650	Note 2
	48	100 X 100	390 X 650	Note 2
PLCC	44	070 X 070	400 X 400	
	68	105 X 105	700 X 700	Note 2
	84	140 X 140	900 X 900	Note 2
PQFP (Bumpered)	84	115 X 115	400 X 400	Note 1
	100	150 X 150	500 X 500	
	132	225 X 225	700 X 700	Note 2
	164	260 X 260	900 X 900	Note 1 & 2
	196	320 X 320	1000 X 1000	Note 1 & 2
QFP (EIAJ)	44	050 X 050	200 X 200	
	64	080 X 080	350 X 350	
	100	160 X 160	350 X 580	Note 2
	160	280 X 280	850 X 850	Note 2

Notes: 1. These packages are currently designed but not tooled at Amkor/Anam facilities.

2. Mechanical stress is a major concern on large die sizes. Amkor/Anam has no reliability data on die sizes larger than 500 mils.

Package estimates courtsey of Glen Koscal, Amkor Electroincs, and John Heacox, Amkor Electronics.

in every package where the logic allowed in the device may be far less than the minimum die size.

When this situation occurs, logic design changes to a speed versus parallel tradeoff. In college classes the same topic is discussed, but in general, the consideration is that the lower the gate count, the better the cost will be. That is not always the case in ASIC designs, especially when bond-pad-limited. The costs for a raw piece of silicon are not always on a gate basis. It costs no more to have 1000 gates than it does to have 2000 gates in a 40-pin, dual in-line package. Figure 5-25 shows a typical serial-versus-parallel tradeoff in system design and the appropriate speed and gate count calculations for two different high-performance CMOS processes. This is a typical logic implementation, as shown with a Karnaugh map with two different reductions. Speed and integrated-circuit area will vary dramatically in the two different implementations of this logic.

Later there will be a discussion of the aspects of packaging and pin count and package-limited die applications. Often designs get into a situation where the silicon area is consumed by the package and not by the logic. In these cases, the

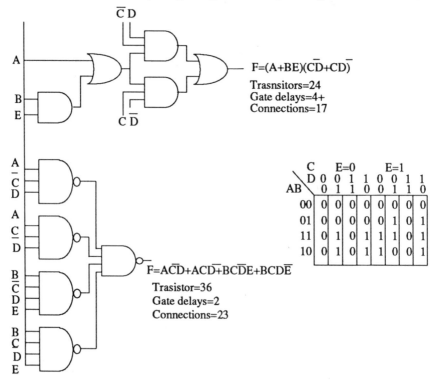

Figure 5-25. Logic optimization.

Table 5-3. Gate count and package types.

PACKAGE TYPE	LEAD COUNT	MINIMUM DIE SIZE, MILS	MAXIMUM DIE SIZE, MILS	3-μm MIN-MAX GATE COUNT	1-μm MIN-MAX GATE COUNT
PDIP	16	030 X 030	145 X 500	225-18125	2025-163125
(300)	24	040 X 040	150 X 500	400-18750	3600-168750
PDIP	24	040 X 040	390 X 600	400-58500	3600-526500
(600)	40	070 X 070	390 X 650	1225-63375	11025-570375
	48	100 X 100	390 X 650	2500-63375	22500-570375
PLCC	44	070 X 070	400 X 400	1225-40000	11025-360000
	68	105 X 105	700 X 700	2756-122500	24806-1102500
	84	140 X 140	900 X 900	4900-202500	44100-1822500
PQFP	84	115 X 115	400 X 400	3306-40000	29756-360000
(Bumpered)	100	150 X 150	500 X 500	5625-62500	50625-562500
	132	225 X 225	700 X 700	12656-122500	113906-1102500
	164	260 X 260	900 X 900	16900-202500	152100-1822500
	196	320 X 320	1000 X 1000	25600-250000	230400-2250000
QFP	44	050 X 050	200 X 200	625-10000	5625-90000
(EIAJ)	64	080 X 080	350 X 350	1600-30625	14400-275625
	100	160 X 160	350 X 580	6400-50750	57600-456750
	160	280 X 280	850 X 850	19600-180625	176400-1625625

Package estimates courtesy of Glen Koscal, Amkor Electroincs, and John Heacox, Amkor Electronics.

designer should look seriously at serial-versus-parallel logic functions. Table 5-3 shows the approximate gate count, along with minimum and maximum die sizes, for multiple different CMOS processes and package combination.

It may be obvious by now that selection of the library could wait until after the initial logic design is done. After package considerations are taken into account, find out if the design is a pad-limited silicon case. If so, add the extra logic to accomplish the task in the same system time requirement, without jeopardizing performance of the system.

DESIGN

Assuming that the tradeoffs have been made and the design is now ready to be implemented, you now must select from the many commercially available tools for the generation of logic in an ASIC environment. The actual process of schematic entry is supported by most vendors on a multitude of workstations, or platforms. The difference between doing a design on inexpensive PC as opposed to a $20,000 workstation is substantial. For small designs, the PC will probably work fine. For larger designs, the capability of an engineering work station for better throughput time and improved features will be a benefit. The design process falls into one of a few categories: (1) straight schematic entry and logic definition, (2) logic synthesis, compilers, data path construction, and other various automated techniques, and finally (3) large block module selection and use. Each will be looked at individually in this section. All these design methods have an impact on testing; that impact will also be discussed.

Schematic entry and logic definition

The implementation of the combinatorial and sequential logic using standard schematic symbols captured on a workstation is a relatively easy task. During the period of generation of the logic and schematic capture, it is very important to keep in mind how this particular block will be tested. How far internal to the circuit will it be when the final chip is constructed?

Although some portions of the logic may appear simple, when they are buried deep inside an integrated circuit, having access to them and being able to test

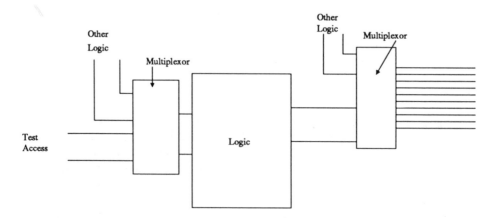

Figure 5-26. Input-output multiplexers for testing.

them may become a difficult task. Remember during this portion of this design that adding extra logic or tapping intermediate nodes to bring them out to multiplexors on the output pins for testability is vital. In addition, if there are multiple input nodes from multiple sources, there is a need to include the ability to force inputs using the same general techniques. Figure 5-26 shows a typical logic diagram with input and output multiplexors in place for increased testability of internal logic.

Logic synthesis

One of the major advantages of logic synthesis and compiler tools is their ability to implement large blocks of logic based on high-level descriptions. VHDL uses the *high-level description language* based on register transfers. The IEEE has adopted it as a standard, and it allows hardware programming in a software type of language. It consists essentially of registered transfers and block functional design.

These techniques lend themselves to very well defined data paths throughout the part and are normally associated with data buses, address buses, and control buses internal to large logic blocks. They are also well-suited to random or combinatorial logic design, such as would be found in the control section of a microprocessor. This data path structure poses problems for testing, but again can be addressed if the access observability and controllability is well-thought-out before implementation of logic. Again, high-level description of this system with access to internal parts makes the task of testability significantly easier.

During the process of simulation several aspects must be discussed relative to testing. First, most simulation done by a system designer will be looking at performance-related characteristics of the device and functionality, that is, propagation delay through gates or propagation delay from one sequential element to another. The designer will compare these characteristics with system cycle time, setup and hold times to ensure that the system is stable during all operating conditions, high and low speed operation. The second aspect is that of functionality check. Making sure the logic is correct for the intended function is equally important.

Block selection and its impact on design

Some vendors have large blocks in their library. The models for these blocks, usually supplied by vendors, can be of several types. There is a gate-level-model, which actually represents the individual logic structure used to implement the block. There are behavioral models, which are essentially the outside ring of the device with internal timing and a high-level description language of the internal works of the device. This model can emulate the functionality from pin to pin and cycle to cycle.

Vendors support the testing of these high-level blocks of logic in multiple methods. One method is to take full responsibility for testing the large blocks. Another method is to turn testing over to the logic designer of the system based on their own internal use or modifications of the blocks. Weigh this very heavily when deciding which vendor to use or not use. For large blocks it is far better to have the vendor choose and support a standard test pattern set.

When selecting macros or large blocks, keep in mind how will they be accessed and tested. Now is the time to add logic to improve the testing of these blocks if the vendor does not. Does the vendor need special access to get to the blocks of logic? Does the vender do test patterns by themselves and add them to the final test program?

Other considerations

In addition to strict logic design, there are some physical and electrical considerations involved in the use of blocks. Many vendors will specify blocks with certain control pins brought to the outside of the package or bond pads. There are also form factors of the blocks that will limit what type of package a device can be implemented in.

Multiple fixed blocks

The condition of multiple fixed blocks is yet another challenge to the ASIC designer and foundry. Although it is quite easy using present design techniques to insert several large blocks into a design, getting them to fit is often a problem. The form factor of the blocks, number of blocks, and overall die size (again this may be package-related) all must be taken into account.

Figure 5-27 shows the design of a moderate-sized ASIC with several large blocks. This diagram shows four different blocks implemented in an ASIC with the physical size of the blocks and the problems of interconnect for the blocks. Notice that there are examples of multiple blocks being put together which will not physically fit inside the package. This problem has several potential solutions when implemented in silicon including:

- Rearrangement of the blocks to fit (this may not mean the optimal layout for the device performance) or a less-than-desired layout for density of interconnection between blocks.

- A change in the physical form factor of the blocks. This is especially true if any of the blocks are compiled or comprised from soft macros.

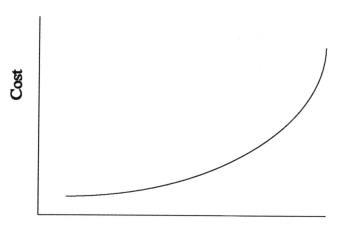

Figure 5-27. Cost versus die size relationship.

- An increase in the package size to a accommodate larger capacity.

- Deletion of some logic or functionality.

None of these choices can be weighted or a decision made until the full problem is understood.

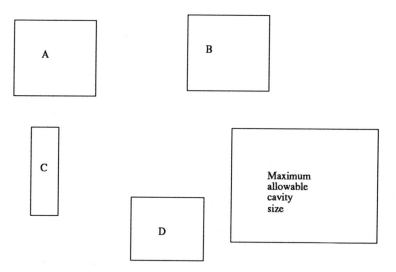

Figure 5-28. Multiple fixed blocks.

Table 5-4. Cost of one or two device designs.

Design option	Package	Die Size	Cost, Each	Total Cost
Single package	132	400	8.04	8.04
Two packages	84	295	4.81	
	100	282	4.33	9.14

Note: Not included in cost: NRE charges, for two designs, PCB changes for extra area, effort needed for partitioning of logic

Based on a vendor's capability and expertise in one area or another, a vendor will recommend a certain direction for the design implementation, which could again turn out to be a major cost impact and decision. Figure 5-28 shows the relationship of die size versus cost on a 2 μm layer-metal process from a fictitious vendor. The matrix put together in Table 5-4 shows the approximate total cost for various options for the implementation of the ASIC device correlated with gate count, package, and die size considerations. This is essentially the same as a device-constrained PC board application.

DESIGN OF TEST MODES

Once the design process has been started and individual test modes have been selected, they must be added to the high-level logic diagrams. Next they should be added to functional specification to ensure that the device will be tested in an orderly manner. The test modes must be implemented in a way that ensures they do not adversely affect the overall performance of the device.

It is important to ensure that the incremental logic that is added to critical paths do not force the device into a time-sensitive state. The analysis of the relative logic paths versus the performance of the library should be done at this point to ensure that there is timing margin to add logic to the device. If not, parallel logic or speed-optimized design techniques would probably be used to increase the speed of the system to ensure minimal marginality. Again it is very important to ensure that all the testability is added during the *initial* logic design process. If, in the process of going back during later sections of the design and

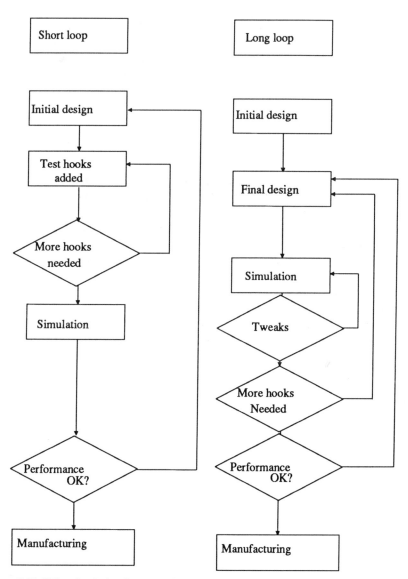

Figure 5-29. Effort for design for test at the start.

verifying the device functionally and testability, there is a need to add logic for testing, one may need to go back and resimulate the entire device. This is done in order to ensure that it correctly meets the system specifications. Figure 5-29 shows two different flows for the design using long and short iteration cycles. This figure examines the results of ensuring testability early in the design versus the impact of waiting until later. Paying attention to testability in the early portion of the design will significantly shorten the overall design cycle. The impact on the design cycle for well-thought-out testing is dramatic. If the logic for testability had been implemented in the original design, the original logic simulation would include all the timing and parasitic degradation related to the test logic added to the part.

SUMMARY

This chapter has looked at many different methods for improving testability of parts during the design process. All these methods work, and they work very effectively. Choose multiple ones to implement in the device. The most important aspect here is that if you are working with a large design that is going to be complex, you must make provisions for testability. Doing so will pay off significantly during the portions of system debugging and production test capability, described in the later chapters of this book. It is very important to remember now early in the design cycle, what kind of tester will be used for the device test. The tradeoffs from one tester to another will influence what can be implemented in the logic. As discussed in Chapter 2, in certain cases a tester cannot easily handle all the conditions that a circuit designer can create. You should design for a minimum of tester capability if doing so does not compromise circuit design.

6

DESIGN AND SIMULATION FOR FUNCTIONALITY, PERFORMANCE, AND TESTING

After you have completed the original schematic entry process and logic design, the next major step in the process is simulation for functionality. This step is then followed by simulation for electrical performance of the device and simulation for testing. These simulations may be done in many different ways—for example, by actual delay path simulation through a single path, or by functional analysis of the part on a state-by-state basis, using propagation delay analysis through individual paths to make sure that the performance of the part meets specification. Any of these methods work, and will work correctly, if they are implemented using the right techniques as supported by the vendor.

It is very important to insure that patterns for logic functionality and timing checks, are as comprehensive as possible. Analysis of the device, the logic in the device, and the ability to completely exercise the logic with patterns now become one of the major concerns.

If you assume that the part is laid out with access to the individual nodes for testing purposes, it is a relatively easy process to exercise the device with logic test patterns. This is no different from the original simulation and logic design problems that were presented in most college classes. What this step entails is the generation of ones and zeros to check the logical operation of the device.

SIMULATION WITH THE TESTER IN MIND

When looking at these blocks of logic with the tester in mind, remember the restrictions and the capabilities of the tester that were selected, as discussed in Chapter 2. Items such as length of data patterns and the number of timing generators available in the selected test system should especially be kept in mind during the simulation process. This is to ensure that the logic functions correctly and is not exceeding the resources of the tester for future test.

Take a look at each block of logic and vectors as tests are being generated. Check the test requirements against the resources of the machine that will be needed to test each block. Ensure that all the patterns transfer from the simulation to the testing environment. If the engineer is inexperienced in testing, allow a safety margin by assuming that there are fewer timing generators on the system (timing generators are almost always the constraint). This allows growth for future changes to the patterns. If the patterns do not fit, the earlier it is fixed and corrected, the better. Check the logic functionality to ensure that the tester can correctly exercise the device, so that the task will be easier. Go through and implement the one and zero patterns looking for the proper functionality. Keep in mind that the tester has been selected. AND/OR logic exercising is the same for an ASIC device as for a big system except for the size of the system. This may be in blocks of logic of a hundred to a thousand gates. Of course, the larger the block of logic, the more complex the input stimuli pattern. One rule of thumb is that for every two-input NAND gate equivalent there should be approximately four test vectors or four states of the patterns necessary to exercise the part. Remember, however, this is only a rule of thumb.

In many cases complex designs that have high testability built into them can be tested with a significantly smaller number of data patterns per gate. There are also devices that are very complex and may take significantly more than one vector per logic gate. During the process of simulation, understand what the critical path of the device is and how to measure it. The critical path is the one with the least deviation from specification. It may be the path with the most gates, but not necessarily. Figure 6-1 shows an example of multiple paths through a part and an analysis to identify the critical path. Any of several of paths could be the critical path, and these are denoted in boldface lines. Of course, the critical path is dependent on the sequential elements at both ends of the path and on the length of the combinatorial path. Again, this may not be the path with the highest gate count, but because of varying propagation delay through the different gates in the path, multiple simulations may be necessary to find which path is the most critical within the part.

During the process of functional simulation, look at the actual parameters in which the system will be operating: What were the vendor-supplied delay times for slow and fast performance of the gates within the library? Do the operating conditions used in simulation agree with the proposed system specification for the use of the ASIC device?

Once the simulation is completed, examine the simulation relative to the system's specification to find out if it is necessary to add guardbands. Guardbands are the margin between the simulated or actual performance of the device and the *needed* performance of the device. At this point some circuitry may need to be redesigned to meet the speed or timing goals of the system.

PATTERN GENERATION AND PARTITIONING

During this portion of simulation it is important to realize that the pattern length is defined primarily by the tester. The normal simulation technique is to look only at the functionality of the device within the system. Saving the patterns that are generated during simulation for future testing makes the test program and pattern generation significantly easy to complete. Functional testing can be used as a starting point but rarely makes for good concise testing. However, most users start with functional tests anyway; they at least usually provide a good base for testing.

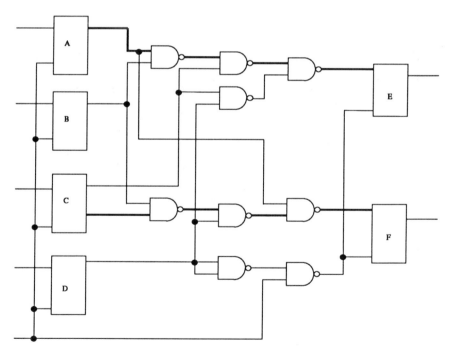

Figure 6-1. Critical paths.

Partitioning patterns into blocks that are identified for individual logic blocks will aid significantly in the debugging of the device. Figure 6-2 shows a logic block diagram for the device with four discrete logic blocks. In this example, the direct-access test capability is added to the blocks to control observability and controllability of the individual blocks. There is also a test mode pin which allows selection of individual blocks for control and observation.

Test patterns could be generated for this block that would exercise all data patterns continuously. This would require the setting up of one block prior to another for the functional checking of the device. In this particular case, if simulation is done without the understanding of the maximum vector depth of the verification system or design system, patterns will need to be redone to ensure that the functionality is checked during the verification and production testing process. If, on the other hand, the test programs and patterns are generated in small increments and identified with individual blocks of logic, the task is easier. This could include patterns that exercise subsets of a larger block to facilitate debugging of the device.

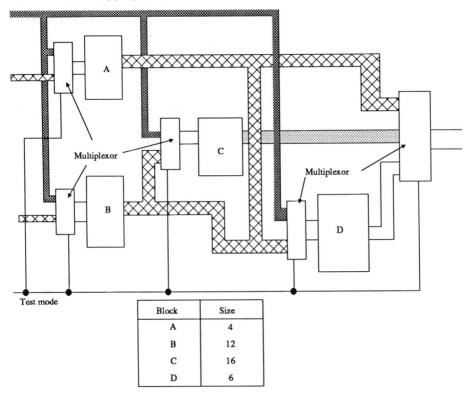

Block	Size
A	4
B	12
C	16
D	6

Figure 6-2. Testability added to four blocks.

A sequence of generation of patterns would be like that shown in Figure 6-3. This shows the partitioning of patterns into individual blocks. If these patterns were generated with the verification system in mind, they do not need to be repartitioned for testing after the initial simulation is done. In this particular example, the test mode should be exercised first to ensure that the reset and test mode works. The test mode checks the condition of all the parts and logic blocks that are read out directly to the pins of the device to ensure that the test mode is working correctly. This is done prior to implementation of any test. In Figure 6-3, the four blocks of the device are checked on output pins during reset. They are checked individually through the multiplexors to ensure that the test mode

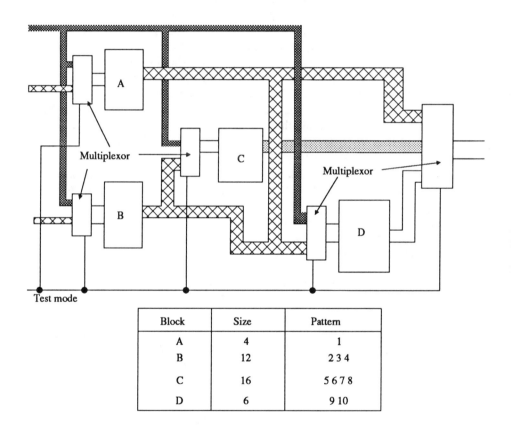

Block	Size	Pattern
A	4	1
B	12	2 3 4
C	16	5 6 7 8
D	6	9 10

Figure 6-3. Partitioning of patterns for blocks.

logic works. This verifies that the block selection register allows proper storage of data, that the multiplexors are connected correctly, and that the I/O pads associated with them are functional.

In the selection of which block to test first for logic functionality and verification, start with the smallest, simplest block that contains the minimal amount of transistors, then gradually tackle the more complex blocks. In this particular case, block B is the simplest block. It contains the least amount of gates, and can be tested without partitioning of the data patterns. To test this block, force the test mode logic into the selection mode for block B. Data patterns for block B and clocks would be supplied to the inputs of the part, and block B would be observed directly on the outputs of the part.

Once the patterns are completed, another block would be selected. In the case of block D, the patterns required may be greater than the 4K limit on some logic verification testers. This particular block contains 10,000 gates of logic, and the vector rule of thumb says that 10,000 gates requires 40,000 vectors. This means that for a 4000-vector limit in a test program, you should partition the testing into a minimum of 10 discrete sections. Each section should be independent, with all the setup and control conditions present within the logic block test.

It is possible to generate the logic block testing sequences based on the setup of previous conditions including the states of previous outputs of the part, but it is a very bad idea to do so since it is very difficult to debug the device when trying to look at one block by itself. To analyze failures, it is wise to reset the device and ensure that it is in a known state prior to execution of data patterns for a particular block in every vector file. This also saves fault simulation time. Looping on a test and checking outputs with a scope or logic analyzer is a method for understanding test failure of a part. If the patterns require previous setup, the entire program may need to be included in the loop. This means that the oscilloscope or logic analyzer may be able to capture data only once every second or two. This makes it hard to see a 1 or 2 ns problem.

If the data patterns are generated so that they are stand-alone, such as continuous data patterns that do not require the setup of a previous block, testers could then easily cycle between the beginning of the test pattern and the first failure. The engineer responsible for the device could look at inputs and outputs with a logic analyzer or oscilloscope and use other engineering tools to find out what is functioning correctly and incorrectly within the block.

Guardbanding

Guardbanding is the difference between the worst-case propagation delay and the minimum required for the system spec. In general, if there is excess margin or guardbands have been used, the device yield will be more stable and have higher quality. Figure 6-4 shows the typical guardbanding of a signal in the system. This particular logic chain is shown with its fast and slow simulations and the value needed for the strobe to ensure the slowest-possible simulation. The delta between the strobe and the actual slow value is the guardband.

Some of this guardbanding is used for testing as the difference between tester measured value and the actual specification of the part. This additional guardband capability is a very important aspect and should be consciously looked at by the designer.

Figure 6-5 shows the product distribution in a typical manufacturing process. This distribution holds for virtually every parameter in the process and likewise for all ac and dc specifications of the part. If one of those parameters is important to the design and critical in the system, check to see if there is guardbanding in the parameter.

In Figure 6-6 notice the actual distribution of the parameter relative to the processing of the device in the fabrication process. The distribution shows multiple points on the slow end of the distribution. Those points are the test values for production testing, the test values for Quality Assurance (QA), the system specifications, and the actual need relative to other components on the PC board. In the example shown in Figure 6-7 there is guardbanding between each of these parameters, and the guardband is in the wrong direction. This figure shows the incorrect placement of QA value, test value, and specification value.

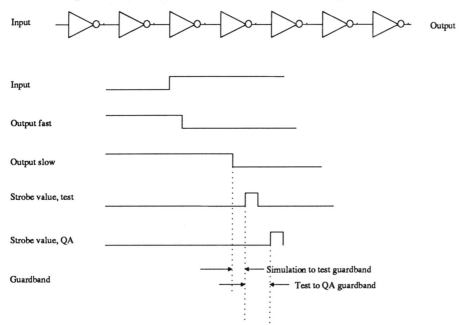

Figure 6-4. Guardbands for testing.

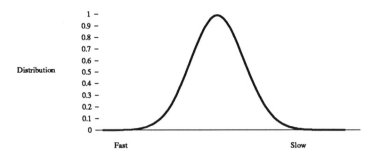

Figure 6-5. Distribution of parameters.

Guardbanding added in the incorrect direction is called *reverse guardband*; it is a major cause of yield and performance problems. If guardbands are not carefully checked, negative guardbands can be generated, which may result in poor yields and a nonmanufacturable part.

Understanding this aspect of guardbanding is very important for ensuring the repeatability of the device that is about to be designed and manufactured. Guardbands should always be looser than needed for the specification of the devices. If only one guardband is requested, it should be the QA guardband. In this case, the specification, QA, and needed value are one and the same, and the margin is minimized.

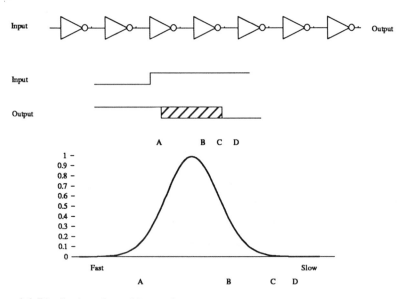

Figure 6-6. Distribution of speed in a path.

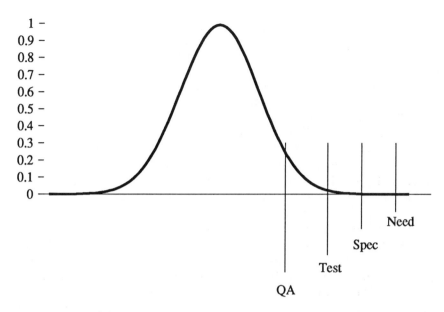

Figure 6-7. Proper guardbands.

FINAL CHIP SIMULATION

After individual blocks are simulated, the blocks are assembled into the chip, then the final chip is simulated. This simulation will ensure that the blocks properly communicate with one another and that the device meets system specifications.

Depending upon the type of system and the performance requirements, it may be necessary to use simulation techniques to look at packaging and interconnections from one device to another. To do so, you may also have to look at performance external to the part to ensure that the system functions accurately. Some items of interest include package lead inductance and capacitance loading. Now is the time to check prelayout performance with estimated capacitance for the correct system specification.

There are simulators that are commercially available which can look at these parasitic. Fault simulation and accelerator engines are also available to look at full chip and multichip system simulations. These become very complex and can consume a significant amount of computer time to implement the simulation in the environment supported by the vendor. This high consumption of computer time means high cost for simulation.

The focus of attention now turns to the process of actual manufacturing of the device. The next steps are related to building the ASIC device.

Physical layout

There is still the ability to vary floor planning and bonding considerations within the device. Assuming that a particular package has been chosen, there is still time to move data buses around. Connecting modules in different ways and partitioning them into different physical sections of the chip can still be done. Depending on the vendor that is being used, the customer may or may not have the access to move modules and do what is called final place and route. One of the steps in place and route is block layout or floor planning. This physical rearranging of the large blocks within the device will impact the performance. When moving blocks of logic around within the device, remember to go back and look at the critical paths and ac performance of the part. Once again ensure that it is meeting system requirements.

Most vendors will have requirements for minimum power and ground connections. They may also have some test pins and other fixed pins that may require accessibility to the outside world. Such items as microprocessor blocks may require the data bus, clocks, and control signals to be brought out directly to the outside pins of the part. This allows the vendor access to the device for testing. Although the customer may have the ability to go in and override it, this is a big mistake because it eliminates the ability of the vendor to test the large logic blocks which are implemented within the system.

Grounds

Selection of the number of grounds and ground pad placements is another important topic that needs to be discussed. Depending on what the ASIC is driving or what the external capacitance is, the design may generate a great deal of noise. The library and high-speed switching of the outputs that were chosen also impact noise.

Compensating may require significantly more grounds than originally planned, when first looking at implementing an ASIC device. The physical ground connections are very important for ensuring that the system meets the V_{OL} and V_{OH} levels as specified. In addition, the buses that are driven that have high capacitive loads with fast switching speeds pose basic problems to a system's designer. The government has specifications for the allowable electrical noise emitted by the electronic system.

Not much can be done to improve the inductance of a package. There is, though, the variation between packages from one family to another. It may be necessary to select another package type to improve inductive responses and noise. There are also techniques such as using separate grounds for input or output bond pads and the internal logic of the device. In addition, slower rise time drivers may be available in the library. One may also solve the noise problem by implementation of logic within the device that controls switching of data bus pins a few nanoseconds apart for each output port. This will minimize the risk of a 32-bit bus switching from all ones to all zeros simultaneously. This can be done only if a margin or guardband is built into the simulation of the part. If there is time in the system to allow data to ripple through all the outputs so that only a few of them are switching at the same time, the problem can be minimized. There is of course the potential of decoupling on the circuit board for further noise reduction. Finally there are circuit techniques that can limit switching time, slowing down drivers but not decreasing drive strength.

In the case of large buses, alternating ground and output pins is another alterative. This additional grounding lowers the noise from both emitted radio frequency and power supply noise. Noise from the power supply, if excessive, can cause logic within the device to switch states.

CUSTOM PACKAGES

One of the final resorts for solving testing, packaging, and noise problems is the ability to generate custom packages. Package changes can vary from simple modifications of currently available present commercial dips, PGAs, and quad flat packs to major changes. Tooling of such a custom package is a fairly long and expensive process. One of the advantages of custom packages is the ability to add internal ground planes and internal V_{CC} planes without adding extra pins. Take for instance, the example shown in Figure 6-8 of a 100-pin PQFP design with four grounds and four V_{CC} connections.

Using an internal plane could easily add as many as four more ground connections and four additional V_{CC} connections to the device without having to add a total of eight additional pins to the outside of the package. This effectively provides for a 108-pin package. By keeping the system pin count relatively low, the bonding capability for the device is relatively high. The die size table cited in Chapter 3 assumed that there were no additional internal planes and extra pads. The assumption for that table was a 100-pin package with a 100-bond pad ASIC device in it. Adding four grounds and four V_{CC} connections increases the minimum die size by the size of the pads. Depending on the layout of the pads this could be from 24 to 32 mils on each side of the die, thus increasing the cost of the die.

Package selection includes not only plastic and ceramic, but also die sales for a chip on board program.

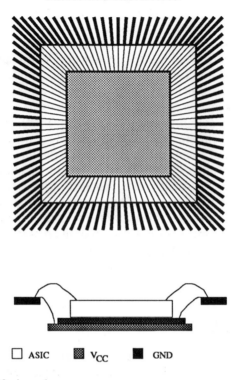

ASIC V_{CC} GND

Figure 6-8. Custom 100-pin package.

SIMULATION FOR TESTING

During the final phases of simulation, attention needs to be turned to the simulation of patterns that will be used for testing the device. Patterns for both production and prototype verification testing must be considered. If the same tester is not being used for both verification and production, two sets of patterns may be needed.

At the outset, be sure to examine the initialization of logic and ensure that everything is in a known state at the time of the start of test. Test patterns are executed on the entire part. This means that if test logic is incorporated into the part to access multiple blocks on some pins in the part, the final device is still done in parallel. There is a need to exercise modules not only in parallel, but also to set up initial conditions for the entire part. This requires the states of the part to be known at the time of test pattern execution. Again, this is the time to look at the tradeoff between additional logic for testing and test time versus total die size.

Time-based simulation

Typical simulations are done on a time-based system. Table 6-1 shows the output of the simulator for a small circuit with the appropriate times and pattern changes cited. This represents the output of a typical simulator. Times listed in the left-hand column are those of the simulator and are not tester-ready. Note that the time axis is not related to T_0 of the tester or to a selected set of reference time sets. Inputs to the timing simulation are done whenever the engineer who is generating the patterns selects the time for an input transition. Outputs occur after the logic has propagated the function to the pins of the device. Most simulators will keep track of input, output, float, and perhaps unknown states.

As was discussed relative to testers, there are limitations of cycle time and input stimuli placement in the tester cycle on the test system that may require the patterns to be modified. Table 6-2 shows the same data patterns reformatted with the tester in mind. In this particular case, each cycle of 50 ns starts over with a new T_0, and output changes are referenced to that T_0. Timing parameters then become a relationship of the pin, the vector, the format, and timing requirements. Many vendors supply software to make this conversion and to identify problem areas.

Table 6-1. Time-based simulation data.

TIME (ns)	PIN NUMBER																			
	1	2	3	4	5	6	7	8	9	10	11	12	13	14	15	16	17	18	19	20
31	1	0	1	1	1	0	1	1	1	1	1	1	1	1	0	1	1	1	1	1
52	1	0	1	1	1	0	1	1	0	1	1	1	1	1	0	1	1	0	1	1
97	1	0	1	1	1	0	1	1	1	1	1	1	1	1	0	1	1	1	0	0
101	1	0	1	1	1	0	1	1	1	1	1	1	1	1	0	1	1	1	0	0
156	1	0	1	1	1	0	1	1	1	1	1	1	1	1	0	1	1	1	0	0
207	1	0	1	1	1	1	0	1	1	0	1	1	0	0	0	0	0	0	0	0
229	1	0	1	1	1	0	1	1	0	1	1	1	1	1	0	0	0	0	1	1

ADDING TEST CAPABILITY

In the previous section you saw that considerable time was spent on breaking blocks into manageable blocks of logic for simulation purposes, then simulating each block, and finally, simulating the full device. Assuming that the patterns were saved, one pattern that could be run in the final simulation for testing all these patterns. This would allow access to the block as they are simulated and would use the same patterns that were used for simulation of the blocks for the testing of the individual blocks of the device.

This practice of adding logic for test capability is an area that will have an enormous impact on the design and testability of the part. Again, this all assumes that there is available space on the die to add test capability, that it can be done without having to increase the die size beyond the maximum allowable in this particular package and without having to convert to a higher-density process, and finally that it is not a critical path in the device.

Table 6-2. Tester-cycle-based simulation data

	TIME (ns)	PIN NUMBER
		1 2 3 4 5 6 7 8 9 1 1 1 1 1 1 1 1 1 1 2
		0 1 2 3 4 5 6 7 8 9 0
	31	1 0 1 1 1 0 1 1 1 1 1 1 1 1 0 1 1 1 1 1
	52	1 0 1 1 1 0 1 1 0 1 1 1 1 0 1 1 0 1 1
	97	1 0 1 1 1 0 1 1 1 1 1 1 1 0 1 1 1 0 0
	101	1 0 1 1 1 0 1 1 1 1 1 1 1 0 1 1 1 0 0
	156	1 0 1 1 1 0 1 1 1 1 1 1 1 0 1 1 1 0 0
	207	1 0 1 1 1 1 0 1 1 0 1 1 0 0 0 0 0 0 0
	229	1 0 1 1 1 0 1 1 0 1 1 1 1 0 0 0 0 1 1

Vector	Time	Data	
1	0	1 0 1 1 1 0 1 1 1 1 1 1 1 1 0 1 1 1 1 1	
2	50	1 0 1 1 1 0 1 1 0 1 1 1 1 0 1 1 0 1 1	N = NRZ
3	100	1 0 1 1 1 0 1 1 1 1 1 1 1 0 1 1 1 0 0	R = RZ
4	150	1 0 1 1 1 0 1 1 1 1 1 1 1 0 1 1 1 0 0	1 = R1
5	200	1 0 1 1 1 1 0 1 1 0 1 1 0 0 0 0 0 0 0	
Format		N N N N N R 1 1 1 1 N N 1 1 R N N 1 N N	

If the design is at the boundary of die size and performance where one of those two options needs to be decided, that is the tradeoff and test capability and functionality of the part or different design techniques must be used. Remember the intent of doing this is to ensure that the device is manufacturable and testable.

Resets

For the combination of combinatorial and sequential logic, remember to add a reset pin. This can be a single pin controlled by a test mode, or it can be directly accessible. The reset pin actually sets the state of every storage element within the device to a known state, allowing the immediate exercising of data patterns without long routines necessary to reset the device. Figure 6-9 shows a logic circuit implemented with a controlling state reset signal. This shows circuitry for the single reset pin that is allowed to control all the sequential elements within the device. When testing a part, it is very important to ensure that all storage elements are in a known state. The counterargument takes into consideration costs, not only in terms of in area but in speed degradation in the performance of the ASIC device.

Alternative resets

Yet another possibility for the generation of resets for control of individual states in the part for testing purposes is to use one of the test modes of the part to select and implement a reset. Using the example of a shift register to supply test modes to the part, one of the states of the register could be a reset signal. This would be temporarily turned on and turned off during the course of testing. This figure uses a test mode and a clock into a shift register to be decoded as a reset for the device. See Figure 6-10 for an example of one such circuit.

The problem with this particular method is that there may be a situation where resetting the device during the test mode sequence does not perform the same function as a reset during the system application. Another way of stating the problem is that in the real world, the system may not have access to this particular reset mode. Therefore, the device may go into spurious states. This is no way to ensure that it is in the proper functional state for the logic and operation that is expected of the part.

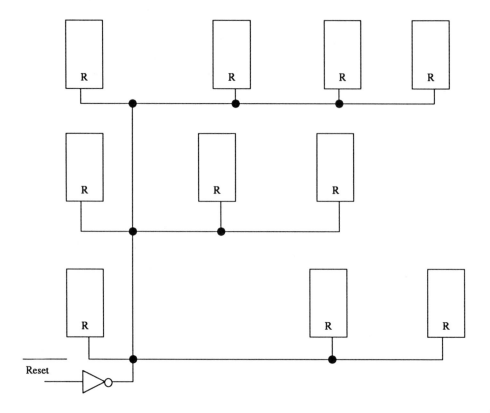

Figure 6-9. Common resets.

Test tradeoffs

There is a tradeoff between logic (both in terms of area and time delay) for testing and test time. Perhaps, the easiest example to understand is one where long counters are involved. For example, watch or clock circuitry requires the use of test modes. In Figure 6-11, there is implementation of a digital logic clock circuitry using standard logic techniques and no special test mode capability. Notice that some of the elements take a significant amount of time to cycle. No special features are added for test capability and the approximate test time is shown in the figure. This is not a practical situation. The input frequency may be relatively slow during normal operations, and if it were increased for testing, the impact is still a long test time for the functionality of the part.

Figure 6-12 shows the same circuitry with some extra test logic. The approximate test time for each of the blocks is shown. It is important to note that test time is money for most manufacturers and they do not want to tie up

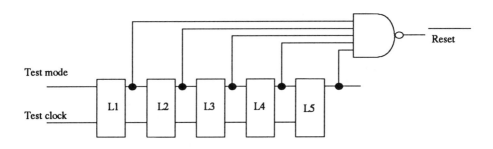

Figure 6-10. Alternative reset method.

valuable tester resources unnecessarily. Usually keeping to a few seconds of total test time for vectors is an acceptable pattern length. Most ASIC vendors would like test programs for vectors to be in the few-second range and will charge extra for longer test of the device. Check with the ASIC vendor if there are restrictions on the maximum number of vectors or time for vector execution to avoid incurring additional costs per unit.

If you compare Figures 6-11 and 6-12, you will notice a dramatic decrease in test time. This testability will have a big impact on the cost of testing of the device and the ability of the manufacturer to make it. In addition to that, based on the observability of the part, now is the time to add extra logic to control states and to make extra nodes observable. Adding logic to the circuit design to ensure that the devices can be easily observed from the outside world is a sure-fire method of ensuring that debugging can be achieved on the part once it comes back in silicon form.

One important assumption that can be made now is that the device will not work, and you are wondering what is wrong as you look at an oscilloscope. If the ability to go back and add logic gates existed, what would they be? Where would the complex portions of the circuitry be? If there is a complex decoding scheme in the state machine, perhaps you should add capability to look at not only the inputs but the outputs that directly control sequencing. Or look at methods to force a "know state" for testing.

If there are control registers or critical paths throughout the part that need to be precise in the timing relationship to one another, adding logic at this time to observe the nodes is a fairly easy task to perform.

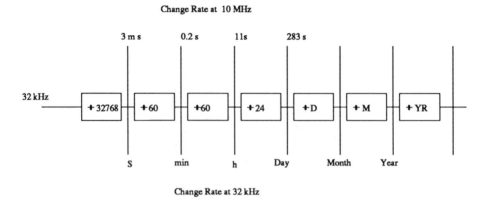

Figure 6-11. Clock circuit with no testability.

TESTABILITY OF A STATE TIME MACHINE

Controlling and observing the state sequences of a state time machine is no different from the combinatorial and clock examples shown earlier in the chapter. Figure 6-13 shows a state diagram of a simple machine with five states, two input variables (A and B), and the appropriate logic to allow transfer from one state to another. The normal operation of this block of logic requires state-to-state transfers that are controlled by a combination of previous states and the decode logic. The checking of this state transition table during normal testing may be a difficult task. Forcing the machine into a given state and setting the

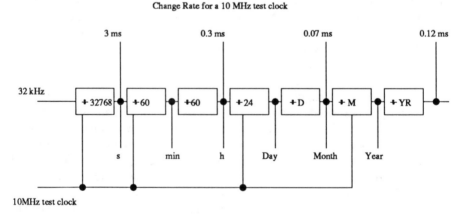

Figure 6-12. Clock circuit with testability added.

proper instruction can be time-consuming. In addition, the three illegal states cannot be checked without test mode logic.

Figure 6-14 shows logic implementation of the states of the machine. Only the latch inputs and control logic are shown for simplicity. The three flip-flops are labeled for bit significance.

Assuming that the machine is controlled by a data storage element (RAM or ROM), an easy method of adding test capability would be to direct-access and monitor the outputs of the gates that drive the state generators. Thus, access to node C in Figure 6-15 would allow the override of the signals controlling the state. This is an acceptable method for checking the resultant operation of the circuit for that state, but it does not ensure that the logic is operating correctly. Control and observation of the nodes is far better. Another place for control of the circuit would be to use locations A, B, and D to preset conditions, and then to clock a single transition and observe the results. This method checks the logic, and the state storage elements for proper operation, but adds more circuitry. The decision to use location A, B, C, or D in the circuit should be made based on margin to specification. If there is sufficient timing margin in the circuit, controlling points C and D while observing the outputs gives the best test. The use of only points A and B allows for less logic in the critical path, but the test

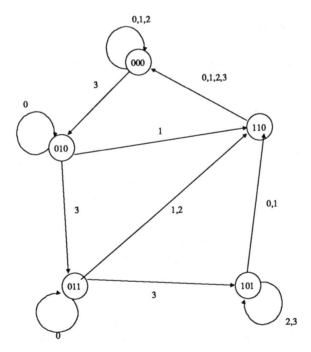

Figure 6-13. State diagram.

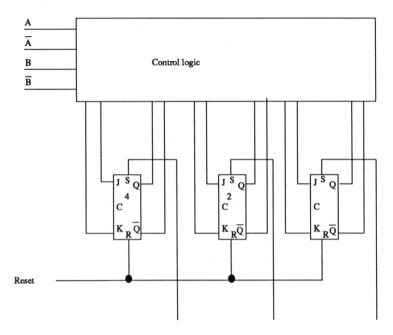

Figure 6-14. Test logic for a state time machine.

sequence may be significantly longer and more complex to ensure the full operation of the circuit.

Figure 6-15 shows the previous figure where modifications for state controllability and observability would be added to the circuit. The test sequence would be as follows:

1. Present the proper address to the part to allow decoding of a specific instruction. Then, based on the known state of the device, monitor that all the inputs to the state flip-flops are in the proper condition. This is to be applied on inputs A and B, or the inputs to the J-K flip-flops.

2. Allow the machine to clock one cycle, and ensure that the state transitions were done correctly, again observing that the outputs of all the state generators and storage flip-flops are in their proper condition.

3. Repeat this sequence for all states of the device. This sequence then allows the direct observation of the data on the output pins of the device, shown as an E on the drawing, assuring that the sequence the machine is going through is correct for the conditions of the initial state of the machine and for the instruction presented to the machine.

Illegal states in a state time machine

This is the time to check the device for illegal states. What happens when a spurious state is encountered? Good design practices will anticipate this and allow for the resumption of correct state time operation. The ability to use test modes to force illegal states and ensure they return to normal operation and not wind up in a loop is a valuable test. Since there are adverse noise conditions in all integrated circuits, the device may go into an improper state. Testing that the machine is in a loop that is not escapable is easy to do with this test technique.

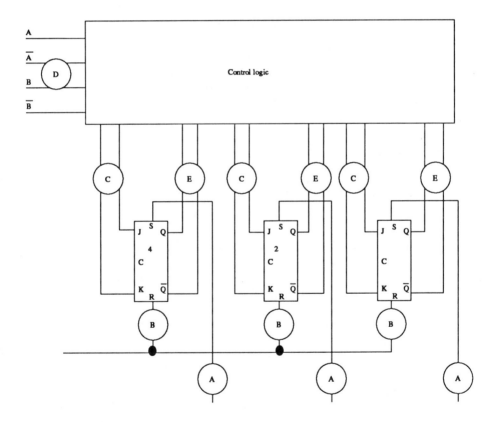

Figure 6-15. State machine with testability added.

ARRAY ELEMENT TESTING

RAM, ROM, and PLA testing are yet another challenge for device pattern generation. In many cases vendors will supply stand-alone patterns for the testing of RAMs, ROMs, and PLAs. Of course, ROM and PLA data patterns require the actual programming of the device to be implemented in the test pattern, which point to the need for a direct-access test.

Some types of RAMs and ROMs have failure modes that are worthy of discussion at this time. In most cases the generation of PLA and ROM patterns is relatively simple. To exercise the part, assume that a simple binary count pattern for addresses will be sufficient for the testing of the ROMs and PLAs. Of course for large PLAs with 100 or more inputs or terms, that may not be practical.

When RAMs are designed, they can be implemented with either static or dynamic circuity. Self-included refresh circuitry may be added or not added if they are dynamic RAMs. Dynamic RAMs are often chosen in large designs due to the much smaller cell size, array size, and power consideration. Figure 6-16 shows the cell of a typical static and dynamic RAM that could be implemented in an ASIC device. Notice the typical configurations of a six-, three-, and a one-transistor RAM storage cell. The use of dynamic RAMs in ASIC devices is scarce, due mainly to the difficulty of testing them.

Due to the higher density of dynamic RAMs, certain interactions take place that need to be discussed. Hopefully, testing for this interaction will be taken care of by the vendor in a canned program, especially if the device is speed-sensitive.

Figure 6-17 shows a typical RAM layout for a multibit RAM; this figure shows the detailed layout of a 64K-bit static RAM. Address decoders, the array, and sense amplifiers are denoted on the figure. The rows and columns are laid out to ensure the best density. A simple pattern of reading and writing zeros and ones into the RAM, although it shows the individual bit significance and isolation, does not tell about row, column, or bit interaction. Patterns such as read-modify-write, marching ones, marching zeros, checkerboard, and galpat (galloping pattern) are good patterns for the exercising of RAMs. In addition, some devices have known sensitive-to-failure modes, such as the adjacent bit capacitive coupling. This may be a reason to use some of the complex patterns such as column and row disturb patterns.

Figure 6-18 shows a galpat pattern implemented for a small 16 bit RAM. This pattern is an N^2 pattern, meaning that for 16 bits, it goes through 16^2, or 256, possible transitions.

Essentially bit A is written with a one while the entire RAM has zeros as content for their cells. The address necessary to read the RAM is switched between bit A and then every other bit; it jumps back and forth between A and B, then A and C, then A and D, and so on, exercising in the entire array. This

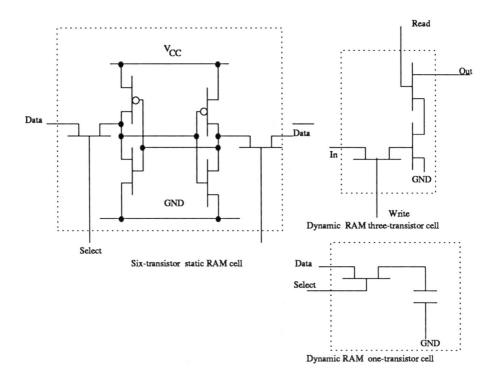

Six-transistor static RAM cell

Dynamic RAM three-transistor cell

Dynamic RAM one-transistor cell

Figure 6-16. Basic memory cells.

allows the complete testing of all sequences of data and interaction of data. Once bit A is tested, bit B is then tested with a one where all other cells are held at a zero. This continues until all bits are tested individually with a one, while the rest of the array is held at zero. This pattern becomes fairly long and involved and, if not run at a high speed, will take a considerable amount of time. This is a good test for all address transitions.

A read-modify-write pattern writes the array to all zeros, then places a one in bit 1 of the device. The full array is then read. Next bits 1 and 2 are written with a one and the full array is read. This continues until bit n -1 and bit n are written with ones and the array is read. This sequence may be repeated for a starting array of all ones.

If the RAM is buried in the device and not directly accessible, major test time could be needed. Figure 6-19 shows a RAM accessible directly through multiplexor logic to the outside pins of the part and the impact on test time if it is not accessible. This figure shows the effect of the RAM testing on a buried and a not

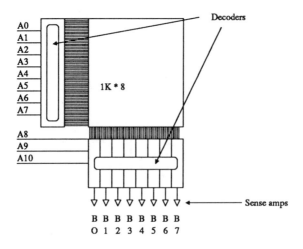

Figure 6-17. A 64K RAM layout.

buried RAM using an N^2 pattern. One of them takes 160 million transitions, the other is a simple 16 million.

Figure 6-20 shows a similar pattern for a column-disturb type of pattern. This particular type of pattern is one where adjacent capacitive coupling from one column to another affects the ability of a cell to store data. In this test a cell is

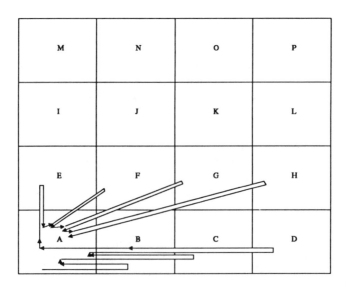

Figure 6-18. Read-modify-write pattern.

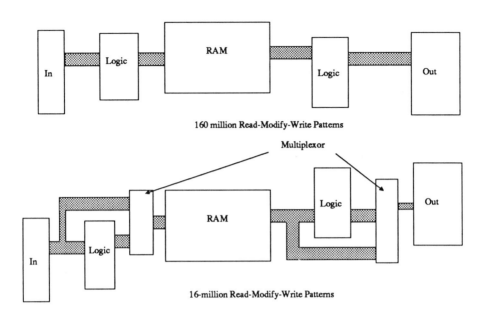

Figure 6-19. Impact of test time on RAM accessibility.

written with one value, be it a one or a zero, depending on which type is the most sensitive in the RAM storage cell. Adjacent columns are exercised and read over and over. After a significant amount of accesses to the adjacent column, the initial bit is reread to ensure that it is remaining in its proper state.

SUMMARY

The generation of patterns for testing, and the design process for testing are not complex problems. The major concern is that elegant and complex designs tend to consume an inordinate amount of vector space and tester resources. Many engineers are usually proud of how complex a block logic they can design for a system. Whereas for testing, the simpler and smaller the design is the better. The other aspect of testing that is important is that long patterns for logic testing of blocks may have a multiplication factor when incorporated in a device. This make design for testing and short patterns desirable.

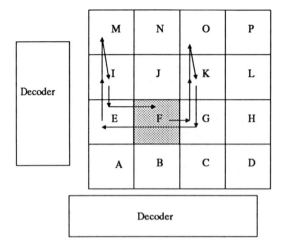

Write	F
Read	G
	K
	O
	K
	G
	E
	I
	M
	I
	E
Read	F

Figure 6-20. Column disturb pattern.

7
REVIEW AND CONVERSION
OF TEST PATTERNS

Now that the patterns are almost complete, your effort must focus on enhancing the patterns for good fault coverage while avoiding extremely long test patterns. This needs to be done prior to net list submission for prototype generation because logic changes may again be needed to ensure good fault coverage. Given a set of input stimuli for a given circuit, how can you tell whether they are good? What needs to be done to make the fault coverage of the patterns acceptable? Monitoring of faults detection by input stimulus in a device can take many forms. These various forms will be covered in the chapter, along with the costs and problems associated with each method.

In addition to enhancing the patterns, this chapter will focus on the grading of patterns for faults, along with methods for checking ac performance of the device. Once again, this may require logic changes to make sure the goals of a good highly testable device are met.

FAULT GRADING

Fault grading is a measure of test coverage of a given set of stimuli for a given circuit. This figure is used to ensure that the device is fully testable and measurable. Once the test patterns are developed for the device, artificial faults are put in via simulation to ensure that the part is testable and observable.

Figure 7-1 shows a single stuck-at-one fault injected into logic circuitry with data patterns for checking the functionality. Notice in the truth table that faults on one gate show up on an output pin of the part or test observation point. Gaining high fault coverage is a very desirable method for ensuring that the

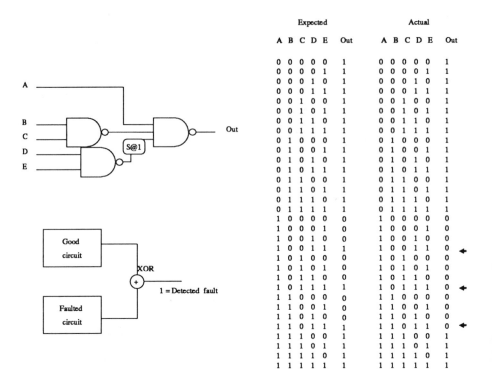

Figure 7-1. Stuck-at-one fault pattern.

device will function correctly in the system. Semiconductor failure modes include not only stuck-at zeros and ones but several other types.

It is also important to mention that although fault grading using single stuck-at-zero and single stuck-at-one faults covers many aspects of semiconductor failures, it does not cover all of them. For example, bridging faults are one of the more common types of failures within semiconductor devices; they may be caught with fault graded patterns if the patterns are done correctly.

One example that is worthy of mentioning is data buses. Assume that you have a design with an 8-bit data bus running throughout the device and in the particular data bus, bits 2 and 3 are shorted together. If, while in the process of testing the data bus, the patterns are simply exercising and observing the devices, an 8-bit data bus could be exercised with a 00H, FFH data pattern. In theory this will cover all cases, and a fault grade program would read 100 percent coverage, but the program will not detect many faults. If bits 2 and 3 are

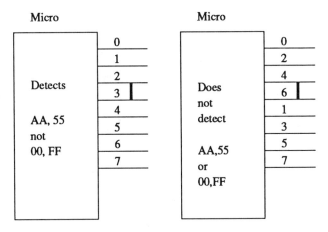

Figure 7-2. Layout-dependent faults.

shorted together, this will still have the failures that pass the test and the fault graded patterns; that is not the case for the pattern of an AA_H 55_H for the same 8-bit bus, for instance. The situation becomes even more complex when the device is physically laid out. Although data bus 0-7 might be in a known sequence on the schematics, it may not always relate to the physical layout of the part.

It would not be uncommon for the pins to be scrambled internal to the part and have a layout similar to that shown in Figure 7-2. This figure shows a microprocessor in two configurations, one with a linear bit sequence and the other with a random bit sequence. Adjacent bit shorts would not be detected with some data patterns. Now is the time to ask the manufacture if the bits will be in order, otherwise do all 256 combinations of data patterns. Even if the bus is laid out in order, taps to the bus may be random. Therefore, pay close attention to the patterns and layout of the device.

One correlation that is valuable is that of faults versus yield of the process versus the defects in the shipped devices. Figure 7-3 shows the relationship between faults and defects per million (DPM). This study shows the relationship of vendor outgoing or customer incoming yield versus fault coverage. It shows the importance of high fault coverage for a good quality level. This study was presented by Harrison et al. at Wescon in 1980. It is included in the proceedings of the Wescon professional program and has been cited frequently since then.

Fault grading issues

The fault simulation types that are described here are not meant to be a replacement for good functional testing. Patterns generated for the functional testing, whether they are functional, random, or pseudo-random generated patterns, are the primary method for testing the device. Fault grading gives only the metric of how many faults are caught for any particular set of input stimuli regardless of the generation technique. Whether they are random or functional, a fault coverage percent improvement means fewer shipped defective devices. Therefore, if the fault coverage with functional patterns is at 80 percent and additional patterns added to the device for testing purposes improves the fault coverage to 90 percent, the testing of the device will be better.

Comparing the fault coverage of 80 percent based on functional pattern generation versus a fault coverage of 80 percent using random or pseudo-random number pattern generation should not be compared. The fault types caught in the exercise of the logic are different in the two methods. Do not mix the two when doing comparisons.

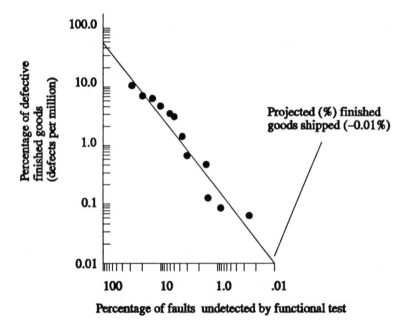

Courtesy of Proceedings of the IEEE professional program, September 1980.

Figure 7-3. Fault grade versus DPM.

Non-stuck-at semiconductor defects

In the case of fault grading relative to defects that cannot be modeled by a single stuck-at model, there are no commercially available tools that monitor and isolate these types of faults. Generating patterns for them has been extremely difficult. Some research has been done in the academic world to find acceptable methods, and although such models exist, they are very expensive.

Non-stuck-at faults can be categorized into several types. They include leakage between adjacent lines, shorts between adjacent lines, leakage to ground, and leakage to V_{CC}. Also narrowing of internal lines in the device and resistive connections cause a longer resistor capacitor (RC) time constant, therefore increasing the propagation delay.

Most of these faults show up as timing or power changes in the part. There have been studies done and some commercial test equipment has the ability to monitor I_{CC} of the device during pattern cycles. A signature analysis of the currents in the device can be generated for testing. Figure 7-4 shows a block diagram of a large circuit with clocks generated at a relatively slow rate and the appropriate I_{CC} waveform. This represents the switching currents for a good and bad device, which is strictly a CMOS characteristic. It is built on the premise of the inherent low quiescent power of CMOS. Figure 7-5 shows the bad device with a model of the leakage current to ground and another model with a leakage current to V_{CC}. Either of these two leakages would show up as a quiescent

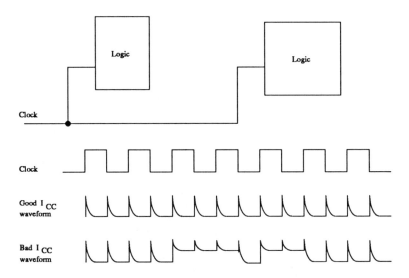

Figure 7-4. I_{CC} waveform for a node with leakage.

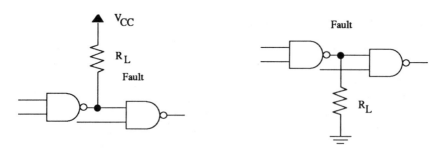

Figure 7-5. Typical leakage fault.

current increase in certain states of the part; this increase is shown at the bottom of Figure 7-4.

Propagation delays show up as slow switching times, which mean an increase in current during the transition periods of the device. Figure 7-6 shows the transfer characteristics of an invertor and the current generated during the transition. For slow transition times, the current generated has a larger total area, and therefore a higher average current value would be seen by the tester.

This relationship of propagation delay and switching current may not show up as a failure in the device. In some cases if there is sufficient margin to spec, the device will perform perfectly well. The current consumption will be slightly

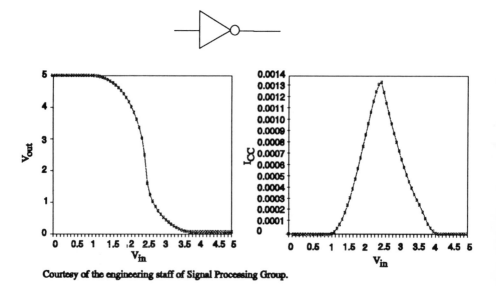

Courtesy of the engineering staff of Signal Processing Group.

Figure 7-6. Transfer characteristics of an inverter.

above normal. In addition, the incremental shift of one gate in a propagation-delayed path is probably within the normal distributions of devices and would be very difficult to screen out relative to the performance of the part.

Figure 7-7 shows the typical distributions of ac performance for a critical path. It also shows the same distribution with one gate with a leakage path to V_{SS} that slows down the transition of that gate by 50 percent. With minor leakages, the impact is very small. A 50 percent difference in the propagation delay of one gate in a chain of nine is effectively impossible to differentiate on a commercial test system. Remember that the propagation delay through that path at the slowest processing and lowest V_{CC} is probably much worse than the typical operation of the device. Therefore, when one path degrades in a typical test condition, it would not exceed the numbers in the test specifications. Chapter 3 discussed the effects of slow and fast processing, which is usually much greater than a 50 percent difference.

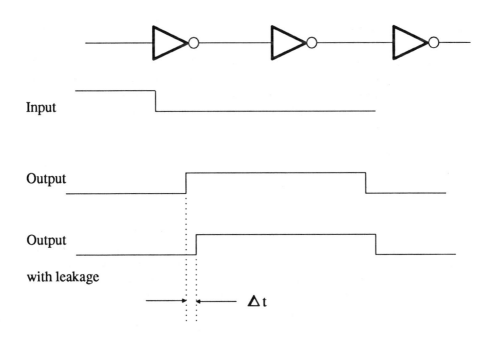

Figure 7-7. Speed impact of leakage.

Toggle count as a fault metric

Toggle count is the simplest method, and also the cheapest method to determine if the nodes are controlled. Unfortunately toggle count only monitors the controllability of the node from an input stimulus standpoint. Devices with good toggle count may have a high number of defects that are not caught in the manufacturing process. Many vendors require a minimum 100 percent toggle coverage as a minimum for acceptance of the design. This pushes the problems of poor fault coverage to the user who must justify the need for better testing and test grading.

One aspect of toggle count that is very attractive is the low cost. Many simulators perform toggle count as the simulator is running, and so the cost is, in essence, free. The other case of probabilistic and deterministic fault grading usually requires separate simulation. Toggle count sets the maximum number for fault grading that can be achieved with the other fault grading methods. This means that if the toggle count is only 80 percent, the highest number that can be achieved with the other methods is also 80 percent. Therefore having a 100 percent toggle count prior to starting the other methods saves time and money.

Probabilistic fault grading

The next step up in fault coverage is probabilistic fault grading. This is a metric where the probability of a fault propagating to the outside of the device is assigned by analysis of the circuit and analysis of the data patterns for testing. In the case of probabilistic fault grading, the nodes are checked to see if they are observable; this does not mean that the faults propagate to the pins of the part.

When comparing toggle count to probabilistic fault grading, the latter is by far a better metric.

Deterministic output fault grading

Deterministic fault grading is the actual presentation of a fault on a particular node within the device. The next step is to check that node to ensure that it is observable at the outside pins of the device. This takes care of two important questions: first, is the node controllable, and secondly, is it observable on the outside? Deterministic fault grading is the best commercial technique for checking faults within the device and providing the ability to observe them.

Output fault grading or grading for outputs assumes that the faults are placed only on the outputs of the driving node. Figure 7-8 shows a typical combinatorial circuit with an output fault noted at the output of gate A. A pattern is

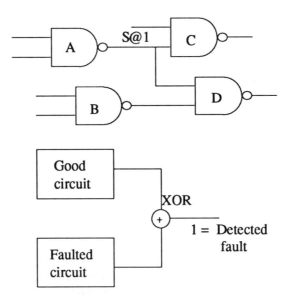

Figure 7-8. Typical stuck-at-one fault.

then generated, and any of the paths that lead to the outside world could check output A to ensure that it is observable from the outside.

Just because the output node is observable, it does not mean that it is covered by all the possible faults. In output fault grading there are several types of semiconductor failure modes that would not propagate through to the output of the device. Figure 7-9 shows an open line with a line floating that reflects as a one on the device. In this case, the output node, as shown in the figure, that is tied to a one could be propagated to the outside causing the pattern to pass. This figure shows a stuck-at-one output fault impressed on a circuit and an open line. The output fault could be propagated through and shown on the output E and observed. In actuality this circuit is bad because the lower portion is not working correctly for a second reason due to the fact that one of the input paths does not work. If the stuck-at-one fault did not exist, the patterns would pass this device.

Input fault grading is the next best case and consumes the highest amount of simulation time. Input fault generation assumes that each possible input of the device has the ability to be stuck at a one or stuck at a zero. It checks the patterns for the ability to propagate input stuck-at conditions to the output of the device. Looking at Figure 7-10, all possible input faults are denoted on the circuit, and patterns would need to be generated to check each possible path. Input faults are shown as numbers. The combination of all those paths versus the patterns that are generated determines the relative fault grade percentage. A 100 percent

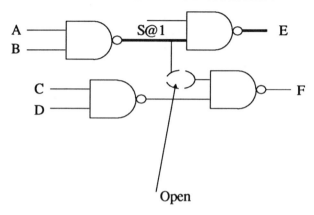

Figure 7-9. Problems with output faults.

coverage means that all possible paths are checked. Figure 7-10 shows all input and output faults. Inputs are numbers, output faults are letters.

The state of the art today is to perform both input and output fault grading. That way both input nodes and output nodes are faulted and the circuit is graded relative to the test pattern for ensuring good test coverage.

Indeterminate faults

Note that some conditions within the logic of the device that could never exist might show up as faults for fault grading purposes. These false conditions need to be analyzed to ensure that the cases are not relevant and need not be checked. Examples of indeterminate cases are redundant logic, multiple states of a state time machine, decoding structures, and other circuit structures.

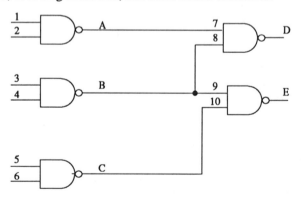

Figure 7-10. Input and output faults.

COST OF FAULT GRADING

Of course, the cost of fault grading is related to the type and techniques used. The cost of doing a simple toggle analysis is by far the cheapest. Statistical or probabilistic fault grading is the next more expensive. Deterministic output fault grading is still more expensive and accurate. Deterministic input fault grading is still more expensive and accurate. Finally the most expensive and accurate is a combination of input and output deterministic fault grading. Then, again, the accuracy of the models and the accuracy of the number generated for fault measurement is better in the more complex cases.

Faults versus yields and DPM

One important aspect in the analysis of fault grading and defects per million of devices shipped, is the aspect of yield. In a paper presented by Dr. Vishwani Agrawal at the IEEE Solid State Circuits Conference in 1982, Dr. Agrawal constructed a model showing the impact of high and yield low processes. This was done with various different fault coverages, and the impact of defects of the final device shipped was analyzed. Figure 7-11 shows the fault coverage required for a field reject rate of 1/1000 based on varying different yielding processes. When this data is analyzed, the conclusion can be reached that a poor

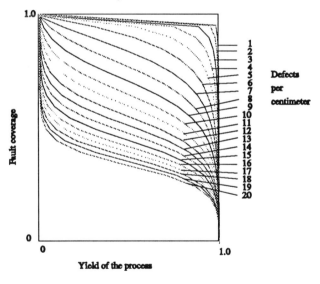

Courtesy of the IEEE Journal of Solid State Physics.

Figure 7-11. Fault coverage and DPM relationship.

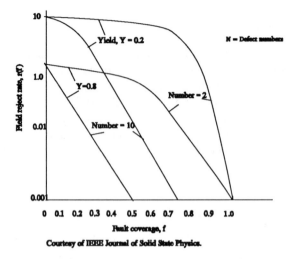

Courtesy of IEEE Journal of Solid State Physics.

Figure 7-12. Relative yield and DPM.

yielding process with moderate fault coverage may get a lower defects per million number than a high yielding process with a higher fault coverage. Figure 7-12 is another interpretation of this data, shown as relative yield and defect numbers. Most ASIC processes yield well, and so fault coverage becomes even more critical.

Table 7-1. Yield data for DPM analysis.

Results of chip test:	Yield $\cong 0.07$	Total number of chips = 277.
FAULT COVERAGE, %	CUMULATIVE NUMBER OF CHIPS FAILED	CUMULATIVE FRACTION OF CHIPS FAILED
5	113	0.41
8	134	0.48
10	144	0.52
15	186	0.67
20	209	0.75
30	226	0.82
36	242	0.87
45	251	0.91
50	256	0.92
65	257	0.93

Courtsey of IEEE Journal of Solid State Physics.

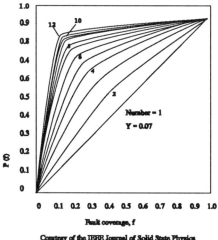

Figure 7-13. Cumulative failures.

Table 7-1 shows the results of chip testing with yield of approximately 7 percent with 277 devices tested. Based on fault coverage of 5 to 65 percent, accumulative failure analysis was done, as shown in Figure 7-13. This data was all generated based on a simple stuck-at fault model. The next section will look at the impact of bridging and leakage on fault models. The important aspect to remember is not only that the fault coverage used for the device test generation routines impacts the yield and the defect levels, but that the yields and defect levels interrelate with the fault coverage to impact the overall quality of the device as seen by the customer.

For large integrated circuits, testing then becomes an even more significant problem. This becomes evident if related back to the control and observation aspects discussed earlier in this chapter. If one does not pay attention to the testing algorithms and the test coverage, large integrated circuits may virtually be impossible to screen and test effectively. (Portions reprinted with permission from *IEEE Journal of Solid State Physics*, 1982.)

FAULT GRADING DISCUSSION

Dr. John M. Acken, for his dissertation at Stanford University in 1988, did extensive work on modeling of failure modes of CMOS integrated circuits. This work focused on the areas discussed in the last few sections, primarily that of actual failure modes of MOS versus the stuck-at model that is primarily used for fault grading for MOS devices. His analysis took several steps. First, a prob-

Table 7-2. Failure types.

FAULT TYPES	NUMBER OF DEFECTS	PERCENTAGE OF FAULTS
Line stuck-at	132	28
Transistor stuck-at	70	15
Floating line	101	21
Bridging	144	30
Miscellaneous	29	6
Total	476	100

Courtesy of Dr. John M. Acken, Stanford Ph.D. Thesis, 1988

ability analysis was done based on types of failures and expected failure modes. In this analysis (shown in Table 7-2) the most likely or most common failure modes are the bridging type. Stuck-at faults, (both the transistor stuck-at and line stuck) add up to 43 percent. Bridging faults are approximately 30 percent of the available faults.

The next step in the process was to do an analysis of the operation of logic with a bridging fault. For this model, Dr. Acken used a voting model based on the relative drive strengths of the transistors of the two gates involved in the bridging situation. To verify this model, Dr. Acken constructed a 6K-bit RAM made up of 24 bits * 256 words. Nine batches of the RAM were fabricated, and Table 7-3 shows the distribution of failure modes for the RAM based on the testing that was completed. Also shown in the table is the distribution of typical

Table 7-3. Failure distribution in devices.

CLASSSES OF FAULTY BEHAVIOR	NUMBER OF CHIPS
Single and multiple stuck-at	28
Complex shorting behavior	4
Time dependent values	2
High V_{DD} current	15
Failed all test	9
Failed extra test tapes	15
Multiple fault types	36

Courtesy of Dr. John M. Acken, Stanford University, Ph.D. Thesis, 1988

failure modes of integrated circuits based on Dr. Acken's surveys of various integrated-circuit fabrication facilities. This distribution shows that the 6K RAM used for the model construction is very close to that of typical commercial devices.

After fabrication, the devices were tested to three different types of data patterns for the analysis of failures. The first data pattern was the simplest possible one, checking only the functionality of the bit. Next a more complex data pattern was used for checking. Finally a very extensive testing pattern was used for the checking of RAM failure bits. The relative failure types of the different data patterns are shown in Figure 7-14.

The voting model

Dr. Acken constructed a voting model in which gates that are tied together fall into five different areas of operation. The first area is the dominant mode where either one of the two gates that are driving the circuit has significantly higher drive strength than the other gates. In this case the output of the high-strength gate always wins. Thus, these are two of the possible conditions based on the two possible gate drive levels.

The next two areas of operation involve drive strengths which are roughly matched, and which may be modeled as either a wired-AND or a wired-OR circuit. The wired-AND is the situation where a zero dominates the output states of the part. For example, the output drive transistors for the low state possess significantly higher driver capability than the pull-up transistors. The opposite state holds true for the wired-OR state, in which the P-channel pull-up transistors are the high power devices. Many CMOS libraries have logic functions that are not designed for balanced outputs; therefore the logic zero may have significantly more drive strength than the logic one. This points to the need for the wired-OR or the wired-AND model.

The fifth area of operation of the model is the state in which everything is balanced. In this case the output is a function not only of the relative transistor sizes but also of the inputs on the gates whose output are shorted. Another way to represent this is to consider the gain of the stages. Figure 7-15 shows the transfer characteristic of a single gate and of two cascaded gates. Note that the area of input that results in an indeterminate output is small for a single gate; for two gates it is virtually zero. The conclusion that can be reached is that any bridging or intermediate voltage on a node will propagate to the outputs as a one or zero in almost all cases. The most likely case for propagation of a midlevel value is the case where the actual output transistor inputs are at an intermediate level. In this case, this will fail a V_{OL} or V_{OH} test.

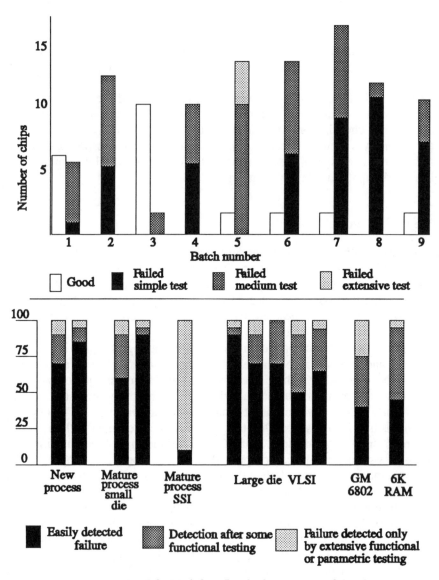

Courtesy of Dr. John M. Acken, Stanford University, Ph.D. Thesis, 1988

Figure 7-14. RAM yield and fallout distribution.

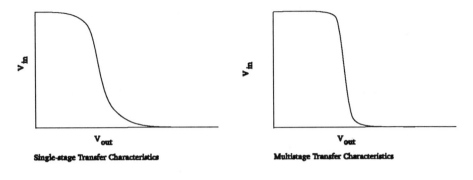

Single-stage Transfer Characteristics   Multistage Transfer Characteristics

Courtesy of Dr. John M. Acken, Stanford University, Ph.D. Thesis, 1988

Figure 7-15. Transfer characteristics of inverters.

Potential bridging faults

An analysis was run using adjacent geometry checking routines. These routines are similar to the ones that are used for capacitance extraction from layer to layer and adjacent lines within MOS circuitry. From this was derived a model of potential line-to-line short sites. These potential short sites are shown in Figure 7-16. The average number of shorts per node potential was four.

Layout considerations for potential short sites

The final section of Dr. Acken's dissertation was on the impact of layout sensitivity on device potential short sites. As stated in earlier sections of this chapter, physical layout varies with the place and route algorithms. Test methods must be dependent on the physical layout for the proper identification and determination of faults that could cause operation malfunction within the integrated circuits. Dr. Acken looked at the logic layout of an arithmetic logic unit that had been implemented using CMOS standard cells and worked with multiple different layout algorithms in order to minimize the logic that had a high number of potential bridging faults. By minimizing the layout and clustering the logic to include a minimal amount of area interaction, the physical design can be laid out in a way to minimize the potential node-to-node shorting problems that may exist in the integrated circuit.

Summary of the stuck-at model

Dr. Acken's model, although representative of modeled faults and hence significantly effective at improving fault coverage for integrated circuits, is not supported in commercial fault grading routines supplied by most semiconductor vendors at this time. One of the important aspects of the conclusion of the paper is that fault grading based simply on a stuck-at model is not sufficient to fully guarantee functionality of the device. In addition, not only can bridging take place between adjacent nodes that are physically close to one another on the integrated circuit, but because of the place and route algorithms used, lines may be connected by bridging across relatively large areas of circuit schematics.

Until accurate bridging models are available in fault simulation routines that are layout-dependent, the only alternative a designer has when generating test patterns for the integrated circuit is to do sufficient functional exercising of the logic so that the maximum possible number of faults are covered. The other important topics of the thesis are that of the gain of logic gates and the notion that indeterminate values propagate to the outputs as a true one or zero. The conclusion, in any case, is that a combination of good functional testing and a high-level of fault grading will achieve good quality product. (Portions reprinted with permission from Dr. John M. Acken, Ph.D. Thesis, Stanford University, 1988.)

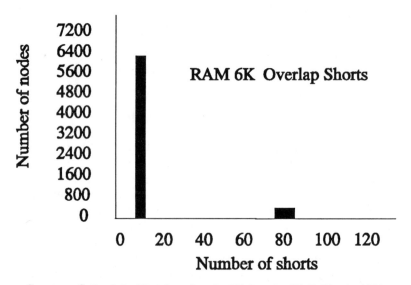

Courtesy of Dr. John M. Acken, Stanford University, Ph.D. Thesis, 1988

Figure 7-16. Shorts per nod found in the 6K RAM.

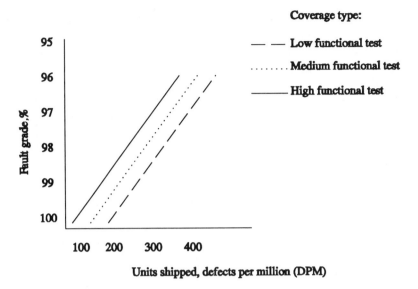

Figure 7-17. Yield versus fault grade DPM.

FUNCTIONAL TESTING

Notice in the work of both Agrawal and Acken on fault grading that there is a hole in the methodology that precludes the total use of fault-generated vectors for good test capability. This functional pattern or functional testing is important to ensure that the defect level is sufficiently low to ensure good system quality.

It is recommended to continue functional patterns on the part and to use as many of the simulations that were used in the functional verification in the testing of the device to improve testability. In the case shown in the previous section, faults such as a shorted data bus effectively move the baseline DPM up. Such faults generate a type of failure that could not be caught without running the functional patterns. This effectively shifts the curves, as shown in Figure 7-17. This figure shows fault grading curves generated with and without functional patterns, which effectively shifts the outgoing DPM up, although the fault grade number is still high.

This increase in DPM poses an interesting problem since it changes the relationship between defect density, baseline DPM or fault coverage, and yield of the process. In this particular case the situation could arise that a low-yielding process and poor fault coverage could actually yield a lower defect rate product than a high-yielding process with high fault coverage. This is due to low-yield processes generally having high defect density. Defects probably impact multiple areas of the device, and so any testing probably throws out most failures.

All of this relates very well to production test program experiences. In sorting wafers and testing units, most of the failures fall into the category of opens and shorts and simple functional tests. This is the simplest possible test that looks only at the inputs and outputs of the device. The actual number of failures that consist only of a single transistor or a single bit in a complex device such as a microprocessor are a very small percentage of the total. This is not meant to advise the abandonment of fault grading; the two go hand in hand. But it is very important to ensure that functional testing is done and that fault grading is not a universal solution to all defect-related problems. Fault grading is a metric of the coverage of test patterns, making certain the patterns test the part in the best possible way.

COMMERCIALLY AVAILABLE TESTERS

As a final note for the simulation of test patterns for the devices, one must pay attention to the final machine that the device will be tested upon. Commercially available automatic test equipment (ATE) used by the semiconductor industry varies dramatically in price and capability. It would be worthwhile to discuss with the vendor what kind of machine they will be using and what kind of restrictions are imposed by that machine.

Most machines have limitations in pin count, speed, timing, edge placement, accuracy of ac and dc measurements, and cycle boundary restrictions. Chapter 2 included a discussion of what these limitations are and how they are used during device testing. Although a particular design may be one that does not run in a sequential or state time mode, all commercially available testers partition the testing into the equivalent of states or cycles. Some testers allow the machine to vary the period from one cycle to another to emulate the asynchronus operation of the system. This way the tester can partition the simulation into tester cycles with the equivalent of T_{-0}. which is the start of any cycle.

In some of the more advanced and expensive machines, timing generators can cross multiple cycle boundaries for a combination of inputs and outputs. In addition, the machines may have variable periods. In all cases there is still some dead time at the end of the final period, be it one or two or more nanoseconds. This dead time allows the machine to reset timing parameters for clock and strobe edges to ensure that the device is exercised correctly.

Simulation for performance and testing

When completing simulation with the tester in mind, you reach a point where race conditions and marginalities in the device relative to testing show up. In addition to that, resimulation with slow and fast processing files with the tester

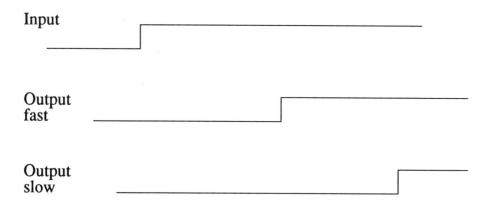

Figure 7-18. Effects of slow and fast processing.

in mind will determine whether the device is fully testable. Figure 7-18 shows the impact of fast and slow simulation and its impact on testing. This figure shows logic with fast and slow processing and two different critical paths. This must also be accounted for when generating test patterns.

Figure 7-19 shows the logic with fast- and slow-process files, and two different critical paths causing test problems. This figure represents the summation of the variables in propagation delay based on processes. The fast representation and the slow representation are the same circuit at the different processing, and temperature, and voltage values of the part.

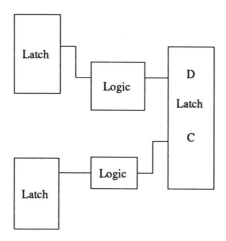

Figure 7-19. Variations in the critical paths.

Figure 7-20 shows the impact on speed of process variations, as a result of voltage and temperature variations. It is important to simulate with the total variation over the entire temperature and voltage range to ensure testability. Remember that if the system design of the logic was not done with some guardbanding, there may be no margin. If the cell library used by the vendor did not include some amount of margin, there may be a need to add guardbanding by optimizing logic for speed or choosing faster gates. There must be a delta placed between the testing parameters used for initial production testing and final quality assurance (QA) test of the devices. This will ensure that they are electrically stable and manufacturable.

Testers, although very accurate and very expensive, do drift, and the temperature conditions on a manufacture's test floor do vary somewhat. Therefore most manufacturers will add a guardband between the specification and the actual test value used for the device.

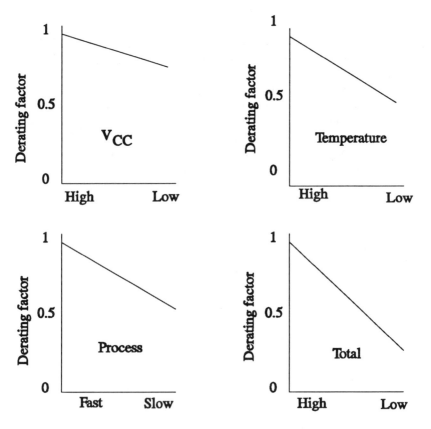

Figure 7-20. Speed variations in devices as a result of process, voltage, and temperature variations.

PATTERNS FOR VERIFICATION AND PATTERNS FOR PRODUCTION

After generating some patterns, one can see the impact of the tester on pattern generation. It will be worth noting that the patterns for verification and debugging may be different from the patterns intended for production test. In the simplest case, the production tester will usually handle significantly longer pattern sets than verification machines do.

When simulating blocks of logic using an initialization sequence, one of the easiest tasks to perform is to break up simulation patterns for use on the eventual verification machine. Start with a small block of logic and create a pattern for that block using only the simulator. This should include all the initialization routines to ensure that the logic is set to a known state. If the verification machine planned for use is limited in vectors, keep the maximum number of vectors in mind as the simulation is done. It is far easier to append several short verification patterns for the production program than it is to split a block of patterns apart. It is also less expensive to fault grade incremental test patterns. Figure 7-21 shows a block level diagram of a microcontroller and the associated reset or initialization sequences. This figures shows a flow diagram of a microcontroller test and the sequence for testing reset, register, and the ALU test. In order to ensure that the patterns fit in the tester, the reset sequence is used

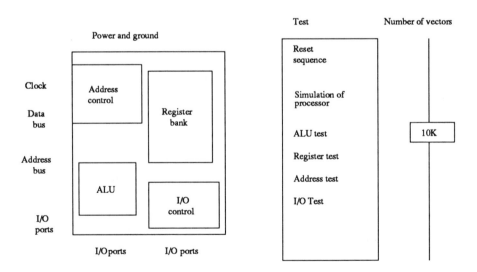

Figure 7-21. Partitioning of data patterns for testing.

at the beginning of all individual blocks being tested and the blocks are kept to 4K or less. These reset patterns could be repeated in every block of the design verification vector sequence.

Figure 7-22 shows the flow for a full vector verification program. In this example the vectors for each block is approximated in the diagram. Notice that a breaking of the patterns means that blocks may not be initialized for test. This figure represents an entire test pattern for a small microcontroller and the partitioning of that pattern into portions for the verification system. If reset sequences are not included in each of the patterns, the patterns may not execute correctly on the system. If you assume that the register test has followed the program counter, you must then be sure it is initialized to run correctly. If this block is to be broken apart, make sure to reset the program counter for simulation so it matches the status of the silicon. If you don't, a lot of time will be spent chasing false errors.

KEEP ALIVE

In certain cases testing of the integrated circuit exceeds the data pattern memory available in the tester. If the patterns cannot be truncated to shorter patterns, there is the alternative of keep alive. *Keep alive* is a feature provided in most test

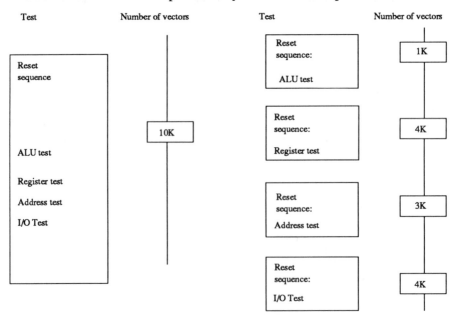

Figure 7-22. Partitioned pattern length.

systems for the continuation of pattern execution beyond the normal limits of vector length. In general, at the end of the data patterns in test systems the tester powers-off the part, thus setting all the nodes in an unknown state in the next power up. This is the reason that all the patterns must contain a reset sequence.

There are two methods of keeping the device alive in the process of reloading data patterns. Most testers will allow the holding of one last vector at the end of the pattern memory execution. All the inputs and power supplies will be held at a stable level while the remainder of the memory is reloaded; this effectively places a long pause on one particular vector, and prevents any changes of state from taking place in the part. It allows for the exchange of data patterns to ensure that the device is fully tested to the length of the data patterns. This technique works only if the device is fully made up of static logic and can tolerate relatively long pauses in the data patterns. Any device that was designed using standard logic elements without dynamic storage nodes or large blocks that include dynamic logic should function correctly with the keep alive of a single vector pause. Figure 7-23 shows the keep alive vector location in vector memory.

The second method of keep alive would involve devices such as some microprocessors which include dynamic nodes and which must have continual

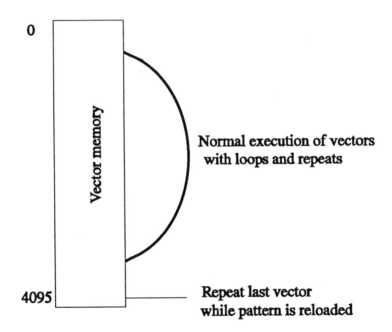

Figure 7-23. Keep-alive single vector.

clocks running to the device in order to ensure that the data patterns internal to the device are not lost. In this case, a loop must be constructed at the end of the data pattern where the tester would loop until the pattern memory is reloaded. To accomplish this, the following sequence takes place.

- Pattern execution of the first portion of the test pattern is completed from vector 0 through vector 4K or the end of the vector patterns.

- A loop is generated of the minimum number of vectors to include timing and data patterns to the part to ensure that the device is clocked and functioning during the reload time.

- The test system controller does the reload of data patterns during the looping.

- A branch condition is put into the loop that allows the tester to notify the pattern controller to exit the loop and start execution again at vector 0. Figure 7-24 shows the looping of vector memory while the patterns are reloaded.

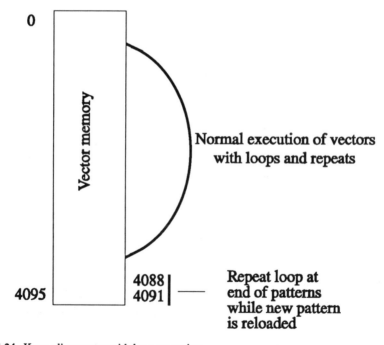

Figure 7-24. Keep-alive vector with loop execution.

The important aspect of keep alive in this second method is the ability of the test system to communicate with the loop being executed. The pattern controller must be able to branch back to the test program for execution of the remaining portion of the test. Keep alive is a somewhat difficult process to program and function correctly for the device. It is an alternative if there is no way to keep the vectors down to a bare minimum that fit inside of the test system memory for proper functionality.

One concern for keep alive with respect to processor type devices is that the typical function that is executed in the loop is usually a nop. The nop is acceptable for not changing data in registers, but usually the address counter is incremented. In this case the first instruction executed of the new patterns should be a jump to a "know" address. This places everything into a know state for testing.

Examples of when you would need keep alive would include checking of long data patterns such as in the error correction code sequences of some devices. You would also use it when testing a RAM if the data patterns were sufficiently long that filling the RAM would require one pass through vector memory for writing and a second pass for reading the RAM. There are numerous other examples.

Simulation for ac performance measurement

When simulation of the device is taking place, one usually targets the intended application for the device. In a typical environment an ASIC device may interface to as few as one other device, in which case the capacitive loading will be small—usually less than 10 pf. In extreme cases such as in microprocessors, capacitive loading may be as high as 200 to 300 pf, thus slowing the part down many nanoseconds.

Testers also impose a load upon the device. Some tester capacitive loads are as low as 30 pf; other machines may be as high 180 to 200 pf. The effects of the capacitive loading on the output transistors of the device is to degrade the rise and fall time and therefore the speed of the device.

When simulating the device for the intended system application, capacitive loads of the approximate size seen in the system should be used in the final simulation. This would ensure guaranteed performance over the range of operation in the intended system environment.

To generate test patterns to measure timing, the capacitance must be simulated with what will be seen in the tester. The effects of adding a 180-pf load to a device that is expected to handle 10 pf of loading can be substantial on the drive capability and timing measurements as seen by the part. Therefore, generating another set of patterns for the ac measurement of critical parameters with the tester capacitive loading in mind is something that will be useful during the design verification and testing phase of the project. To restate, if the path to output under load of 10 pf has a delay of 15 ns in systems use, testing on a machine with 200 pf may require 30 ns. Your vendor can help with this tester-simulation correlation.

Critical path measurements

Not every path needs to be measured and tested within the device. As seen in the previous chapters, some paths will have more margin or guardband than others relative to their simulation for the intended application. After selection of the critical paths for testing, generation of the pattern should be done with a capacitive loading of the system in mind. This capacitive loading and the critical path ac measurements combine to form a rigorous test of the performance of the silicon. This is one method used for determining ac performance of the parts.

SUMMARY

This chapter focused on the quality of the test patterns, not only for fault coverage, but for timing and functional testing. It cannot be emphasized enough how important it is to have good test coverage for the ASIC device prior to submission of net lists to the ASIC vendor. Remember you cannot add logic to the part to help with testing after the silicon has been manufactured.

8

PROTOTYPE GENERATION

Assuming that the patterns have been iterated and that the logic design is stable along with the final device simulation, the next step is to submit the design to the vendor. The vendor will then perform final simulation and net list acceptance. Depending on the vendor's policy and on the performance required by the system the vendor may or may not accept simulations performed on a workstation. Instead, the vendor may require the simulations to be redone on the "golden machine," or reference simulator, back at the factory or design center.

During this step, the simulation patterns that were used for testing and simulation will be rerun on the device. This is called *post layout* simulation, and it is done after the physical layout is completed. The process of the physical design usually takes place with an automatic place and route (APR) system physically placing gates or cells in individual locations. Then the APR system connects or routes the cells together according to the net list. Although the devices are small, typically less than a millimeter on each side, interconnection capacitance can become a dominant factor in the performance of the part, especially one with small geometry parts. Because of this some timings will change.

MOS PERFORMANCE VARIATIONS

The transistors used for MOS device manufacturing for logic gates are typically measured in micrometers, and drive capabilities are quite small. Capacitances in the less than 1-pf level may be sufficient to slow down a gate so that the gate doesn't work correctly in the system. This parasitic capacitance and resistance, associated with the interconnection loads of driver gates and final placement of the gates is what is added to the schematics as delay elements for the final simulation.

If the simulation was guardbanded for performance during the early portion of the design, there may no be need for an iteration of the design. Or if the vendor has good approximation tools, there may not be a problem with timing parameters that no longer meet specifications. Some vendors also allow changes of drive strength of gates, which may compensate for this capacitance load.

It is very important to review the critical paths, timing parameters, and the early simulations versus the timing parameters in the final simulation. The effect of parasitic capacitance and resistance will show up now. This is the time to look back to see whether the performance of the ASIC will still meet all the system specifications.

If the design meets all the device specifications and all the system specifications, now is the time to sign off and accept this simulation. If the simulations do not match the system requirements, this is the time to fix them once again. Most vendors guarantee this simulation as the data sheet for the device to be purchased and do not guarantee anything that was done prior to this simulation. The vendor will not guarantee anything not simulated; so make sure to simulate everything of concern.

If there are errors or timing parameters that have moved too much, talk with the vendor. Ask for certain nets to be changed or to be clustered together or grouped in the routing of the devices. Some vendors allow the customer the capability of doing this on a workstation. Moving cells around to place logic gates closer to one another in the design helps this problem. The APR systems have the ability to move cells from one row or column to another or from one physical location to another. Some of these systems realize what a critical, or important, path is in the part. If the APR system can recognize a critical path, then it can place it with higher priority to keep the nets short. Although it may be possible to do some weighting of the importance of some of the nets, do not do too many. Remember that if all the nets are weighted the same, the effect will defeat the purpose of the weighting.

Other alternatives for fixing slow paths are (1) to increase the parallel function versus the serial logic function in the part, as has been described before, (2) to choose higher drive-strength cells, and (3) to convert to a higher-performance library and process.

This process may be iterative based on what kind of system specs are required. It may be that, if the system runs relatively slow in comparison to the process or if there is sufficient margin, the first pass may be the final one. This is the time to ensure that at the time of sign-off and approval of the electrical simulation, everything is correct for the system specifications. Once the simulation is signed off, the vendor will go through the process of manufacturing the device.

TEST PROGRAM PREPARATION

This portion of the book will look at a combination of methods that can make test program generation easier. This will make the job of the semiconductor vendor, printed circuit board test engineering staff, and the person who will debug the part as easy as possible. This chapter assumes that the patterns have been generated for both design verification and production test as described in the last few chapters.

The vendor will take the data patterns that were presented and turn them into a functional test program. Table 8-1 shows a typical test flow used by a semiconductor vendor for the exercising of the device. It describes the basic test done along with the portions of the specification of the device tested in each step. The flow is from loose testing of the device to detailed checking of specification; this basic flow is both efficient and helpful in debugging failures encountered during test. Table 8-2 shows the approximate test time for a 100-pin ASIC device as executed on a verification system. Notice that a large amount of tests can be done in a very short time. The test times are approximate and will vary based on the type of system used by the type of test executed, length of patterns, and pin count.

The cost of testing is very much related to test time. In production, package handlers, wafer probe equipment, people, and perhaps a computer network are all needed to run production tests. When comparing test time on a small system to a large system, total cost must also be compared. In this case, total cost is the cost of the test equipment along with the needed support equipment. Table 8-3

Table 8-1. Basic test program flow.

TEST	FUNCTION
Shorts	Checks if the adjacent pins are shorted
Open	Checks to see if the p-n junction exits (pad protection device)
Basic function test	Checks functionality of the part, uses most vectors, loose timing and nominal voltages
D C spec test	Checks the inputs and outputs compared to spec for dc levels
A C spec and margin test	Checks the vectors with timing set to spec Checks at minimum and maximum voltage

Table 8-2. Typical test execution on a verification system for a 100 pin device.

TEST	NUMBER OF TEST	TYPE OF TEST	TIME TO EXECUTE
Opens	100	Parametric	1 s
Shorts	100	Parametric	1 s
Basic function test	40,000	Vector pattern	0.04 s*
ac spec test	300-500	Parametric-functional	3-5 s
ac spec and margin test	100,000	Vector pattern	0.1 s*

*Execution time only, vector reload time may be between 10 and 500 s.

shows the same test time for the same device on a large production test system. Again the test time is approximate and will vary depending on options, program flow, and test details. The biggest reason for the difference in test time from a verification system to a production system is due to pattern reload time. As stated in Chapter 2, memory cost is one of the main differences between verification systems and large ATE.

Regardless of the techniques that were used for the generation of the data patterns, the vendor will convert those patterns into a test program. This is similar to that shown in Table 8-4. This is the same flow with the addition of information from the net list and vector simulations submitted by the customer. Vectors are used for functional tests and some portions of I/O testing. Net list description and selection of input pads and power connections are used in the parametric tests. Although the flow may vary from one manufacturer to another, most of them perform essentially the same kinds of tests.

Even if the device used special testing techniques such as JTAG, BIST, BILBO, LSSD, or direct access, the flow is always the same. It is basically always done with a setup sequence or a preconditioning of the part, and a formal measurement of the data on the device under test. The testing may be on a cycle by cycle basis or as a pass or fail conclusion at the end of a long routine. There may be a burst of data to set up the part and a burst of data to measure it. Finally, information is interpreted by the program to categorize or bin the device as a pass or fail rating.

Table 8-3. Test execution on a large test system.

TEST	NUMBER OF TEST	TYPE OF TEST	TIME TO EXECUTE
Open/shorts	2	parametric	0.05 s
Basic funtion test	40,000	Vector pattern	0.04 s
ac spec test	300-500	Parametric-functional	0.5 s
ac spec and margin test	100,000	Vector pattern	0.75 s

SAMPLE MANUFACTURING

The manufacture of the silicon will process the data and manufacture a device in a flow similar to that shown in Figure 8-1. This figure shows the manufacturing flow for an ASIC device from the end of the customer design cycle to the receipt of samples. Steps may vary slightly from one to vendor to another, and there are differences in the length of process steps required for gate array, standard cells, and full-custom devices. The first step is to take the data that was used in APR and turn that into a set of mask plates.

For MOS standard cells, gate arrays, and full-custom devices, this may be as few as three mask layers, or in the more complex CMOS processes it may be as many as 16 to 18 mask layers. Masks can be generated using multiple approaches by either internal mask shops or commercial vendors, and the plates can be generated with a prototype or a production quality specification.

Most vendors using today's processes (3 μm to less than 1 μm) now use either full field projection printer plates or steppers. Steppers can be either reductions of 1* or 5*. The difference between a stepper and a full field plate set is the ability to control fine geometries over the width of a wafer. The expansion coefficients of quarts of low expansion glass must be such that with heating in the printer, the layer-to-layer alignment does not go out of tolerance. This tolerance is usually less than a micrometer, on a 150-mm wafer that is a low expansion rate for full field printing.

With wafer sizes now in the 100- to 150-millimeter range, it is a very simple calculation to determine the allowed run out or expansion characteristic allowed

Table 8-4.Program data sources.

TEST	SOURCE OF DATA
Opens	Process description and net list for I/O pads
Shorts	Process description and net list for I/O pads
Basic function test	Simulations done for test Internal library for cores selected by the user
DC spec test	Net list for types of pads APR data for placement Simulation data for vector setup sequences
AC spec and margin test	Simulations done for test and performance Internal library for cores selected by the user

in the glass. This is to ensure layer-to-layer alignment across the wafer. A 1-μm runout in 150 mm is 1 part in 150,000. This number must be constant regardless of the small variation from machine to machine and temperature changes in the clean room.

Steppers minimize the problem by aligning only a certain portion of the wafer at a time and repeating the alignment over and over. Figure 8-2 shows a full field plate and a stepper plate used in the production process. The cost difference between steppers and full field lithography is substantial, not only in the cost of the plate but in the equipment cost also. In addition, as covered in Chapter 3, the throughput is significantly different between the two. The extra expense for the stepper is sometimes warranted out of need for better resolution and alignment of geometries.

Integrated Circuit Engineering (ICE), in Scottsdale Arizona, has prepared a typical flow of a CMOS wafer manufacturing process (see Table 8-5). This data shows the basic flow of the masking steps for a 2 μm double-layer metal CMOS process.

Other techniques available for the delivery of prototype include E-Beam direct-write on wafers. This replaces mask generation with writing on the wafer directly, thus saving throughput time. Production plates if needed will be purchased later. Cost of the production plates may be pushed to the production run phase.

There are also machines, which are commercially available, that allow implementation of single gate arrays in a very short time (usually hours). These machines modify the metal pattern only, and as of now there are no known tools

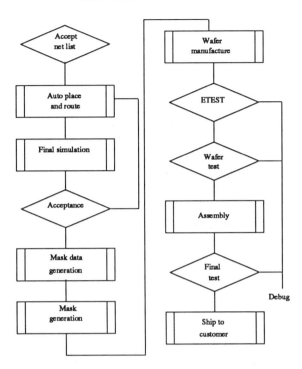

Figure 8-1. Basic manufacturing flow.

to make diffusion variations. The throughput time, or delivery of prototypes, varies dramatically based on processes, vendor, number of masking steps, masking techniques, and quantity.

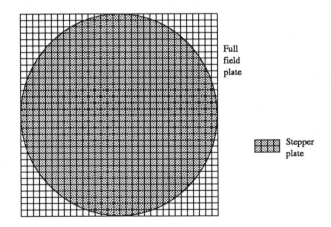

Figure 8-2. Comparison of full field and stepper plates.

After the wafers are fabricated, most vendors perform an electrical test of individual transistors in a test structure. This verifies that threshold voltages, connections, saturation currents, breakdown voltages, resistance, and other parameters are within the specified operating range of the process. Often these structures are in the scribe line of the wafer. Then the wafers may be transferred to an area where the wafer is sorted or tested to the first electrical test program. Here the original patterns generated for testing are merged with a software shell to generate a test program. This implements the program that was described in Table 8-4.

The final piece of equipment needed at this time is a probe card. This is to make contact with the bond pads of the part to the electrical pins of the tester. Most vendors generate probe cards and test programs during the fabrication of the wafer so that all three items come together correctly at the wafer test. Gate arrays typically have a fixed pad placement per array; standard cells almost always need a custom probe card for pad placement.

A test engineer or production operator tests the device to ensure its proper functionality based on the test programs and patterns. This is the first time in the process that electrical testing of the simulation, hardware used for testing, and the test patterns will be checked against the fabricated silicon. The coverage of these items and the simple pass or fail output is a good indication of the test effort put into the design.

Time allotment for a traditional approach of generating mask, fabricating wafers, assembly, and testing throughput time can vary from as little as two weeks to well over ten weeks. This time represents the manufacturing of the device after final acceptance of the simulation.

Note that this is the point when most of the spending of the or nonrecurring engineering (NRE) charges takes place. The generation of the plates, test programs, wafers, probe cards, and packages are the expensive items in the sample process.

The differences in NRE from a 3- to a 1- μm process are basically the cost difference in the wafer manufacturing and plate generation equipment to make those fine geometries on the wafer. In addition, most advanced processes take more masking layers and more processing steps to complete, and therefore, they are more expensive than simpler processes.

Note that most vendors have cancellation policies that may or may not allow recouping some of the cost if a system or design problem is discovered while wafers are in fabrication.

Table 8-6 illustrates a forecast of wafer cost by wafer size. This data is also from ICE and compares the cost of wafers based on 100-, 125-, and 150-mm wafer sizes. Comparable changes in wafer cost are also impacted by the geometry chosen. As you can see, a 1-μm rather than a 3-μm process can impact cost substantially.

Table 8-5. Masking sequence for a 2.0-μm CMOS process

STEP	NAME
1	Well definition
2	Diffusion definition
3	Poly definition
4	N- transistor diffusion
5	P- transistor diffusion
6	Contact
7	Metal 1
8	Via
9	Metal 2
10	Scratch protection

Courtesy of Integrated Circuit Engineering.

PREPARING FOR PROTOTYPES

While the device is with the manufacturer, focus your attention on preparing to test the prototypes when they arrive. This includes wiring the fixtures of the verification machine and preparing the final system for checkout with the ASIC device. It is also the time to convert patterns to the proper tester format and create the test program to debug the part.

Table 8-6. Estimated wafer cost.

WAFER PROCESSING	WAFER SIZE		
COST FACTOR	100 mm	125 mm	150 mm
Facility cost	$75M	$100M	$200M
Raw wafer cost	$10	$18	$30
Depreciation per wafer	$54	$73	$144
Unyielded wafer cost	$123	$150	$233
Typical whole wafer processing yield	90%	87%	85%
Yielded whole wafer cost	$137	$172	$274
Relative cost	0.80	1.00	1.59

* Assumes 80 percent utilization and 5000 wafers per week capacity.
Courtesy of Integrated Circuit Engineering.

Socket wiring

The ASIC device to be tested needs to be placed in a socket on a load board or fixture and wired to the appropriate drivers and power supplies of the test system. The design of the wiring and fixtures for the socket is dependent on the type of simulation, the type of device, and the type of tester being used. These three items must be decided or analyzed at the time of the design of the wiring configuration for the test socket.

The first task is the selection of the power supplies and references for ground and V_{CC}. If the device to be tested uses only one power supply and a ground, select the supply on the tester that most closely approximates the power and voltage requirements of the integrated circuit. In most cases this is a 5-V supply. This supply typically has a range from 0 to 7 or 10 V, and less than a few amps current capability.

Wiring for the V_{CC} and ground supplies should be as short as possible and the connection made at the device under test socket. Some testers have Kelvin capability for the V_{CC} supply and ground connection. In this case, the power and the sense of the power supply can be tied together at the actual device under test pin. This ensures that there is minimal voltage drop on the power line from the output of the power supply to the device under test. For most CMOS devices this will not be an issue, and a few inches of wire will probably not be a problem. For some of the higher-power devices, Kelvin capability is a very accurate feature to have. Figure 8-3 shows the Kelvin connection of power supplies.

When wiring V_{CC} and ground to the device, ensure that all pins that are V_{CC} connections and that all pins which are ground connections are connected with robust wiring. In addition, decoupling should take place as closely as possible to the device under test. This ensures a minimum of noise generation in the test system. High-value capacitors often do not have good high-frequency response. Two capacitors in parallel (large and small) may help.

The actual connections from the device under test to the power supply inside the test system may be significant in length. If there is no decoupling, it is quite easy to generate noise at high frequency of operation of these devices.

Tester signal pin selection for dc drive levels

Partitioning of the test program into V_{IL} and V_{IH} pairs is the next step. On some machines V_{IL} and V_{IH} supplies are independent per pin. On other machines, as noted before, they may be selected in groups of multiple pins. When selecting a particular pin from a tester, it is important to understand what kind of voltage signals will be required to be driven into that pin. For instance, if it is a clock pin or a Schmidt trigger, it may need to be programmed to a V_{IL} and V_{IH} level that is 2 and 3 v, respectively. For TTL levels, typical input levels are 0.8 and 2.0 V.

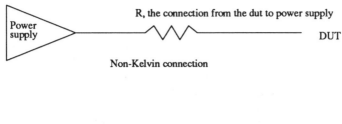

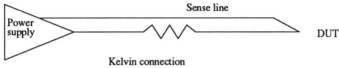

Figure 8-3. Typical Kelvin application.

For CMOS levels, inputs may be programmed to 0 and 5 V. If there is analog circuitry on the device, it will probably need an independent level.

If the machine has these pins configured in blocks by power supply type, it is important to wire the test fixtures with the appropriate mapping of wires. Figure 8-3 shows the relationship of different drivers to different pins on the integrated circuit, allowing the signals and input levels to be programmed for the appropriate pins. This figure also shows the relationship of power supplies versus pins on the part. On a verification system, if the pins are not laid out in an orderly manner on the device, it may require scrambling to have the right voltage levels present for the part.

Tester pin selection for time sets

Time sets also need to be analyzed at the time of wiring the device socket. Some machines limit strobes and input generation time sets to only selected pins within the tester. It is important to understand which pins are exercised by which timing generators. For instance, if a data bus requires the same time for all pins for strobing and formatting of input data, it should be wired in a configuration on the test system where they can all be controlled by the same clock and the same strobe. Figure 8-4 shows the timing relationship of certain pins in the testers and the relationship to the integrated circuit. Specifically, it shows the timing relationship from pins of the part and how they should be grouped together. This is again a consideration when wiring the sockets for the verification system. Different formats may use different time sets to control the formats, adding to the complexity.

All this may seem to imply that one-to-one wiring is not possible. However, machines of sufficient flexibility, there is enough capability within the machine that one-to-one wiring (pin 1 of the tester = pin 1 of the device under test) is possible.

Finally, if resistive loads will be added to the device for the purposes of testing, this is the time to wire the circuitry. As discussed in Chapter 2 regarding resistive loads in lieu of dynamic loads on the test system, it is important to connect the resistive load to the appropriate pin after scrambling of the test signals. The values of R and V will vary based on the loading needed per pin.

A method for allowing the disconnect of those loads during portions of the test to ensure that they do not interfere with other tests being performed should be done. For instance, if the loads are not disconnected, the opens and shorts test, along with I_{ol} I_{oh} test, may falsely fail. Figure 8-5 shows the loads wired with a control switch that allows disconnect by the tester. The signal controlling the disconnect is an extra pin on the tester that can be programmed in the data patterns to turn on and off. This allows control by the tester for each test, if the loads are needed.

After the needs by the pins for voltages, loads, timing, power, and common masking levels are understood, a table can be constructed like the one shown in Table 8-7. This table identifies the groups of the resources in the tester and helps

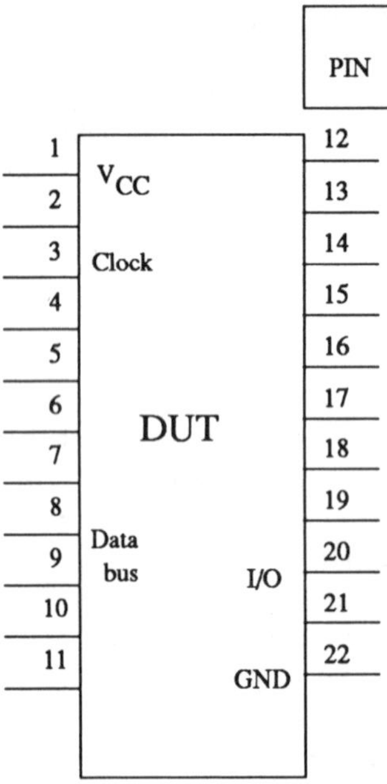

PIN	Signal	Level	Tester pin
12	Clock	CMOS	1
13	Reset	Special	15
14	In	TTL	30
15	Out A	TTL	31
16	Out B	TTL	32
17	I/O C	CMOS	2
18			

Figure 8-4. Unit voltage level selection.

in the allocation of resources to the device under test, by pin type. Based on the restriction of the tester, or the needed grouping of pins, the selection of wiring and connections can take considerable time.

If the restrictions are too great, one alternative is to wire two different socket adapters or load boards to handle different configurations of the device. In this case, some of the functions are tested using one load board, and other functions are tested using another, (see the examples shown at the end of Chapter 2).

Once the wiring arrangement is understood and documented, and the physical scrambling of pins for the device under test versus pins of the tester is completed, actual wiring of the socket and load board arrangement can commence.

Wire selection

Table 8-7 showed the connections needed for test. Other issues of concern during wiring include shielding, impedance matching, wire size and capacitance loading of the device. The wire used should be coax and impedance-matched to the tester (50 to 100 Ω). Capacitance value for the wire, relays, resistors, and pin electronics may be different from that used in simulation. This may mean that small changes in ac values are needed for test.

Improper matching of the impedance of the tester driver and the connection wire is a major cause of noise. Usually the load boards are manufactured to match the driver specification of the test system. Changes in interconnect impedance is best if avoided. If the length of the wire is long and the cycle time is

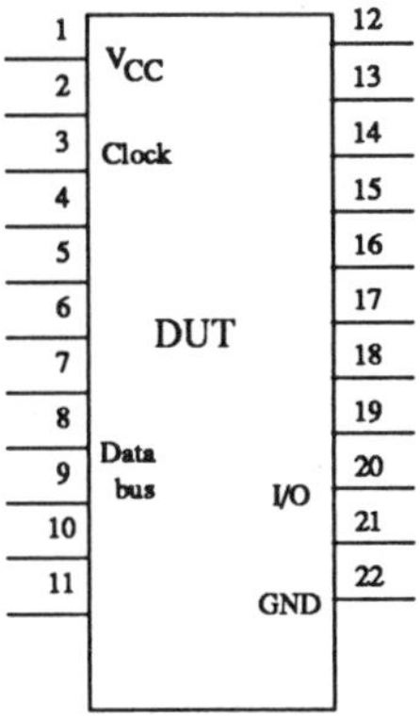

Pin	Signal	Level	Tester pin	Format	Generator
	CLOCK	CMOS	1	RZ	TG 1
	RESET	SPECIAL	15	NRZ	TG 2
	IN	TTL	30	XOR	TG 3
	OUT A	TTL	31	OUT	S1
	OUT B	TTL	32	OUT	S1
	I/O C	CMOS	2	OUT, XOR	S1, TG3

Figure 8-5. Signals and formats for the tester.

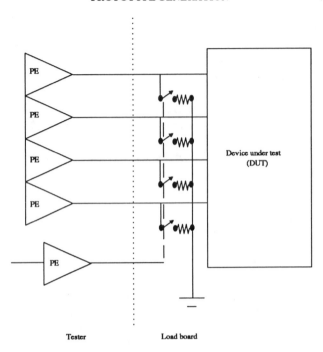

Figure 8-6. Resistive load wiring.

fast, the entire wiring arrangement may need to be reworked. For the cases of very slow signals, mismatches are of less importance.

Socket selection for use on the test system may also impact noise. It may seem strange, but a good-quality socket may cost many times the cost of the ASIC to be tested.

Pattern conversion by pin to the tester channels

This wiring configuration that was just completed will also be used in the tester for mapping simulation signals to tester signals. Test simulation is done in a sequence of pins that were defined by the functionality of the device. Position scrambling of pins takes place in the place and route system and in tester wiring. Make sure the signals are correct. This cannot be overemphasized.

Given the limitations of the tester or the actual layout of the integrated circuit, the sequence of pins in the simulation will probably not relate to the sequence of pins on either the device under test or the tester. This mapping of the simulation signals is essentially a rearrangement of the columns of the simulation to ensure that the patterns are at the right pin at the right time. Table 8-8 shows the simulation of an integrated circuit based on its functionality and the appropriate pattern scrambling to get the right signals to the right pin on the tester. During this time when the devices are being fabricated, it is important not only to generate all of this correctly but to verify that it is correct. A simple scrambling of pins on the tester can cause failures that may take a significant amount of time to understand and to correct.

SIMULATION CONVERSION TO TESTER FORMAT

Now that the wiring is complete, the patterns for the device can be converted into the tester format and a program constructed. Using the configuration map of resources show in Table 8-9, scrambling of the patterns must take place to load the tester. This is a map of the scrambling of pins related to simulation, pin name, signal name, and tester pin. This table should be double- or triple-checked, as there will be problems debugging the part if the pins are scrambled incorrectly. This scrambling can occur using many tools; it can be done by the workstation vendor, by the tester vendor, or by third-party software houses in addition to a simple sheet of paper.

The conversion of other simulator outputs to tester data patterns is treated very much the same. Some simulators output data for unknown states, hi-Z, resistive, sourcing, and undefined. All of these must be matched to the tester

Table 8-7. Pin control map for tester wiring.

DUT PIN	SIGNAL NAME	LEVELS	TIMING	LOAD	POWER	SIGNAL GROUP	TESTER PIN
1	Clock	CMOS	Clock	None	None		3
2	Reset	Special	Loose	None	None	Clock	4
3	V_{CC}	0-6V	None	None	Supply 2	None	1
4	D0	TTL	Bus	4-TTL	None	Bus	2
5	D1	TTL	Bus	4-TTL	None	Bus	5
6	D2	TTL	Bus	4-TTL	None	Bus	6
7	D3	TTL	Bus	4-TTL	None	Bus	7
8	D4	TTL	Bus	4-TTL	None	Bus	8
9	D5	TTL	Bus	4-TTL	None	Bus	9
10	D6	TTL	Bus	4-TTL	None	Bus	11
11	D7	TTL	Bus	4-TTL	None	Bus	10
12	IN	TTL	In	None	None		12
13	Out A	CMOS	S-1	1 CMOS	None		13
14	Out B	CMOS	S-1	2 CMOS	None		14
15	Ground	0	None	None	GND	None	15
16	Out C	Driver	S-2	1K OHM	None		16

Table 8-8. Pin map for the tester.

SIMULATION PIN		DUT PIN	TESTER PIN
13	Clock	1	3
14	Reset	2	4
15	V_{CC}	3	1
1	D0	4	2
2	D1	5	5
3	D2	6	6
4	D3	7	7
5	D4	8	8
6	D5	9	9
7	D6	10	11
8	D7	11	10
9	IN	12	12
10	Out A	13	13
11	Out B	14	14
16	Ground	15	15
12	Out C	16	16

format and drive capability. Testers are limited to drive or not drive, compare one, no compare, or compare zero. Some machines can do compare midband, or float test. In any case, ensure that the data output of the simulator matches the tester pattern requirements. Some of the more advanced tools allow not only the conversion of patterns but the generation of the timing, voltage, and setup conditions for actual testing of the device.

When debugging starts, it is useful to have all this data readily available.

Final interconnection and program check

When the prototype fabrication is complete, it is worthwhile to do a final check of the simulation, tester wiring, program generation, and socket configuration for accuracy. All the scrambling of signals prior to insertion of the prototypes in the socket should be checked again. An ohmmeter is still a good tool for this check.

Table 8-9. Pin scrambling for the tester.

SIMULATION NAME		DUT PIN	TESTER PIN	TESTER NAME
13	Clock	1	3	V_{CC}
14	Reset	2	4	D0
15	V_{CC}	3	1	Clock
1	D0	4	2	Reset
2	D1	5	5	D1
3	D2	6	6	D2
4	D3	7	7	D3
5	D4	8	8	D4
6	D5	9	9	D5
7	D6	10	11	D7
8	D7	11	10	D6
9	IN	12	12	In
10	Out A	13	13	Out A
11	Out B	14	14	Out B
16	Ground	15	15	Ground
12	Out C	16	16	Out C

SUMMARY

This chapter focused on the conversion of patterns and on the preparation for receipt of prototypes. Errors encountered after the final acceptance of prototype net lists from the vendor usually mean extra money spent for more tooling. This step is the real guide as to how compatible the patterns are with the intended tester, and it should not be a problem if simulation for the correct tester was done in the first place. If it was not done properly, there are still some techniques that can be used to enhance the ability to test the part. This will be covered in the next chapter.

9

PROTOTYPE VERIFICATION

When you finally receive the silicon, all the previous work on testability pays off. Once samples have been delivered, several steps can be taken. The first, and simplest, is to plug the device directly into the end system to see whether it functions. If the device is very simple, that may be all that's necessary: Simply plug it into the system and do the necessary routines that check out the functionality of the part in the system you are building.

Assuming that you want more data than a simple yes or no and the design is something relatively complex such as a 10K to a 20K device, you should ask questions about the device's functionality. Do this even if the device functions correctly in the system on the first try. In addition, some vendors may be shipping what is commonly referred to as "cut and go" silicon for prototypes. This means that the vendor has manufactured the wafer, and physically cut it and assembled it into packages. These are wafers that pass the transistor-level E-test for individual die. The vendor then assembles these devices into packages. Based on their process yield, somewhere around half of them may be good, depending on the processing and complexity of the design. Cut-and-go devices are not tested to the vectors that were submitted. In general these types of shipments are decreasing.

Even if the results are bad for the first device that is plugged into the system, try several others. The bad part about plugging devices into the system is that the system may actually damage the ASIC device. This may cause all the prototype devices that were shipped to you by a vendor to be destroyed in the process of analyzing them.

The same potential for damage holds true for debugging the device in the test equipment. Assuming that it is to be tested on a stand-alone test system, one way to proceed is to define a test program and test patterns which are methodical and which would should look at the functionality of the device in ever-increasing steps.

Chapter 10 will describe what is in a commercial production test program and how it flows. Now for the debugging on a bench, a simple routine is worthwhile. The first and most important step in debugging a device is to ensure that all the voltages are connected to the proper pins. This means that you must check not only voltages but the fit of the device into the socket. Therefore it is not exposed to any parameters or high voltage that may destroy the part. Some packages can shift slightly in the socket and cause problems.

HARDWARE VERIFICATION

During the time from net list submission to the receipt of samples, building up the necessary hardware for debugging the device is an activity that can easily consume several weeks. Constructing the hardware for a 100-plus-pin device and verifying that V_{CC} and ground are connected is an easy task. And to verify that pin connections have been done correctly, there is no tool better than the basic ohm meter. Make sure as you check the wire and sockets that you pay attention to the top or bottom view of the device: how will it physically be situated in the socket? If measuring voltages, check also for the proper pins. It is amazing how often a device is wired counterclockwise rather than clockwise for instance.

After having verified all the power supplies and grounds to the device, the next step is to verify that the input signals are driven at the proper time and that the output signals are compared at the proper time. In most design verification systems one can go through the basic steps of setting up the program starting execution and then go through the first several tests to verify that the device is receiving input signals on input pins. This should be done with either a voltmeter or an oscilloscope, and, with no part in the socket.

Prior to plugging the first part into the test system, partition the program into modules that will exercise different portions of the device. If the system is designed in somewhat of a hierarchial or modular manner or if there are areas of logic that are simple, attack those first. Figure 9-1 shows an ASIC system with various different logic blocks partitioned for testing. Isolate the vectors for the first portion, the simplest portion; in this case block D of the figure is chosen. Next attempt to pass patterns to this portion of the device. This exercises the part and look-at functionality of only the simplest portion of the part. Again, this is the point in time when system time and specs become critical.

A typical menu-driven design verification system allows the program to go through numerous activities in the functional test of a single module. In the process of design verification, one of the first parameters to avoid is high-speed testing of the device. Start with the basics and expand upon that. Take for

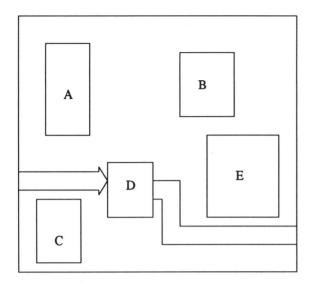

Figure 9-1. Device with logic blocks partitioned for testing.

example the sketch shown in Figure 9-2, and assume that you are building a design verification program for that module. The next few sections will build a basic test for this module.

Continuity check

The first step is to verify that all the pins are properly connected to the part. Most systems will have some kind of pregenerated continuity check. If not, the tester manual will describe how to set up a functional or parametric continuity test. Figure 9-3 shows the characteristics of an input pad arrangement with the diodes for protection and the electrical characteristics of the diode; these are the typical input characteristics of a CMOS device with input protection diodes. Currents are shown for powered-on and powered-off conditions. A continuity test verifies that the diode exists and is functioning correctly; this ensures that the tester is connected to the pins correctly. It checks that V_{CC} and ground are wired correctly and that the inputs and outputs of the device are in contact with the drivers and receivers of the test system. The dual diode may not exist on all pins. For instance, an open drain output and inputs that allow greater than V_{CC} inputs do not have the upper diode.

The basic principle here is that for all CMOS devices there are diodes connected from the input pads or output pads to the substrate or wells. This effectively prevents an unpowered on device from going much above or below 0.7 V

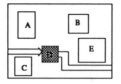

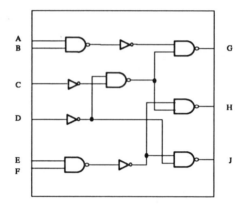

Figure 9-2. Identified block to be tested.

of ground. This test is done by forcing a small current, such as 100 μA, and measuring the voltage with the device powered off. If measuring a high voltage greater than 2 or 3 V, the device is probably open, and there is no physical connection being made to the diode.

If the voltage measured is less than 0.5 V, there is probably a short, which may be some kind of a solder bridge or wiring error in the hardware of the tester loadboard, or the device may itself be bad. This particular test is called an opens-and-shorts test and is very basic to all test programs. It is important to execute this on a design verification system as the pins may be shorted together, which could inadvertently destroy the drivers within the tester. This would cause the machine to fail, and signals could not be driven into the part.

Unless diagnostics are run on a regular basis, it would be difficult to know whether the system is in error, or whether it is damaged. You could waste several hours or days trying to debug the part, where in actuality, the tester is damaged.

Basic function test

Now that the tester has verified that there is device-tester continuity, the next step is to set up the conditions necessary for a small function test. Such a test

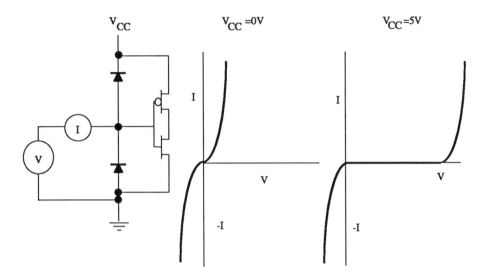

Figure 9-3. Input protection characteristics.

would require a normal flow to select input levels, V_{IL}'s and V_{IH}'s, and output levels, V_{OL} and V_{OH}'s. In addition, input timing parameter strobes and data patterns must be selected. It is best to do only a functionality check of the part. To do so, set the input voltages for outside normal CMOS or TTL operation. This would be 0.0 V for a low and 5.0 V for a high; output levels could switch at 1.5 or 2.0 V, depending on the type of drive capability, TTL or CMOS. For timing parameters, unless the system requires something special, these should be relatively slow: 1 MHz is a reasonable number. Notice in this particular setup that the vectors which would be run are basically looking only at the functionality of the part. These are not measuring any of the timing parameters of the device.

Do not try to measure input and output sensitivity levels along with propagation delay at this early stage. In a new hardware setup with new silicon and new data patterns that have never been debugged, failures may occur. The ability to isolate what is wrong when there is a failure becomes a very difficult task, and so even an experienced engineer takes the test programs that are generated for large semiconductor companies and does what is called a basic function test.

This ensures that the silicon is at least functioning correctly before starting testing of the parameters of the device.

Figure 9-4 shows voltages now set up to drive the part. The V_{CC} ground connections are correct, we have verified all the connections, and input and output from the test system have been verified to the device under test. This has

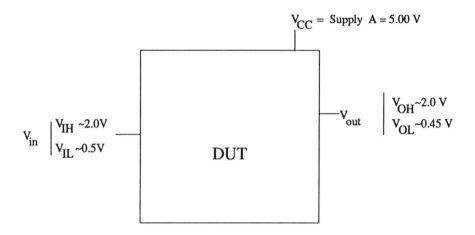

Figure 9-4. Power needs for the DUT.

all the proper conditions in place for testing the device. Now is the time to start data pattern presentation to the part. If a strict combinatorial piece of logic is being tested, do not worry about a reset or setup sequence. To test this block of logic, the follow this procedure:

- All power supplies must be set. In this case we are using nominal or loose values for each supply. V_{CC}, V_{IL}, V_{IH}, V_{OL}, and V_{OH} supplies are needed for the exercise of the part. When switching the supplies in the test program, be sure to pause for the supply to settle. The tester manufacture will recommend a pause value, typically a few milliseconds.

- Cycle times should be set to a loose value. If the system is designed for operation at high speed (20+ MHz) use a 1-μs cycle time for the test. This eliminates race conditions and timing marginalities as a concern in the initial debugging of the device.

- Clock inputs should be at the start of the cycle and set to a minimum of 10 ns. There should be a delay between combinatorinal inputs and clocks that drive flip-flops in the device. This should be set to verify functionality only and not timing conditions.

- Output strobes should be set to the cycle time-minus 20 ns. If a window strobe is used, keep the value to the minimum or at least close to 10 ns. The long delay between input transitions and output delays allows the system noise to stabilize and not effect testing.

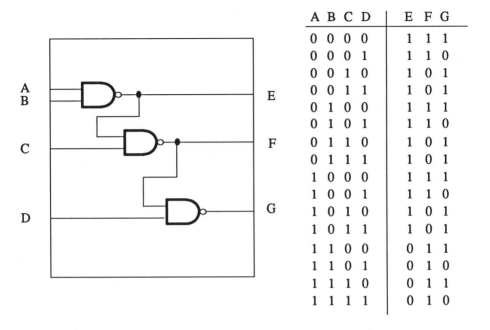

A	B	C	D		E	F	G
0	0	0	0		1	1	1
0	0	0	1		1	1	0
0	0	1	0		1	0	1
0	0	1	1		1	0	1
0	1	0	0		1	1	1
0	1	0	1		1	1	0
0	1	1	0		1	0	1
0	1	1	1		1	0	1
1	0	0	0		1	1	1
1	0	0	1		1	1	0
1	0	1	0		1	0	1
1	0	1	1		1	0	1
1	1	0	0		0	1	1
1	1	0	1		0	1	0
1	1	1	0		0	1	1
1	1	1	1		0	1	0

Figure 9-5. Logic to be tested.

- Formats for inputs must be chosen and supplied to the tester. The last chapter discussed simulation output and the mixing or selection of formats. Format times should not be set to test the margin of specifications of the device during the initial pass of testing.

- Finally, start and stop vector locations need to be specified. This is especially true for the first test attempted. The length of time the tester may need for execution of a few hundred vectors versus many thousands will not be noticeable if the machine does not do reloads. The test time will probably be only a fraction of a second.

Assume a very simple case with only a few gates, as in the example shown in Figure 9-5, debug can start. In that particular case the data patterns used for simulations, the ones and zeros are represented in Figure 9-5. The test program would then take those data patterns, the ones and zeros for inputs, and the ones and zeros for outputs and generate the appropriate timing signals for testing the part. If you assume that the part is made on a normal CMOS process that runs at relatively fast speeds and that it is being tested at a 1-MHz data rate, the data patterns will look very similar to what is shown in Figure 9-6. This illustration

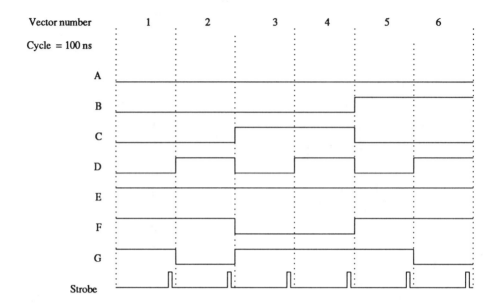

Figure 9-6. Waveforms and strobes.

shows the timing diagrams generated by the data patterns at a 1-MHz rate. Note that there are no marginalities to spec and timing at this speed. The performance of the part is significantly faster than the tester. As viewed on an oscilloscope, the patterns and timing will look perfect.

What we have done here is verify that this simple piece of logic was correctly implemented in silicon for its logic function only. We did not look at the timing parameters. There are several options now. One is to go on to some other patterns or a larger block of logic to test. Another is to try to increase the speed of the machine and the strobes, and start measuring timing parameters to see whether the device is functioning correctly to the ac specification. (Later on this chapter we will discuss the various aspects of parameter testing, but for now, let us continue to look at the larger blocks of logic.)

The next step is to expand to larger blocks of logic. One item that should now be tested is any one of the many test modes of the part (see Figure 9-7). Assuming that there is a pin that allows the setup of a particular condition, again do the same kind of test module generation. First, do an opens-and-shorts test to make sure that the connections to the device are correct. Second, set up a condition to

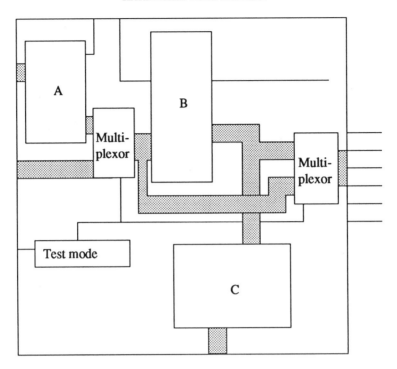

Figure 9-7. Test mode selection.

apply V$_{CC}$ and select input and output voltages and timing, again looking only at functionality. Finally, apply the data patterns as generated by the simulator.

If this block contains test logic, this procedure becomes only slightly more complex than in the previous example. First, set up the logic for one particular condition to be tested. Notice in Figure 9-8 that there is a multiplexor on the output which allows multiple different logic blocks to be multiplexed to a pin for functional checks. During this sequence check the output multiplexor to ensure that all blocks within the device are reset and that their normal reset state is presentable to the output of the device. The first task is to clock in a data sequence which allows the multiplexor to select the logic for input and output purposes that is to be tested. This block is larger than the previous block, but the ideas are basically the same. Go through the vector sequence necessary to test the part, and for simplicity sake assume that it is combinatorial logic again. Here you have several hundred data patterns to the part and measure several hundred outputs. If it is combinatorial logic and it is running at a 1-MHz cycle time, the ac performance of the part is then completely eliminated from concern. There is

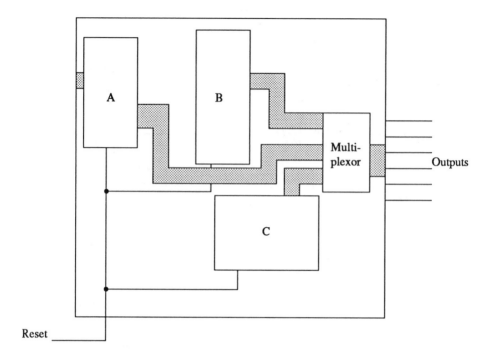

Figure 9-8. Checking the test mode.

no reason to believe that it should have any failure for reasonable lengths of strings of logic.

Even the smallest of automatic equipment will run this pattern faster than it takes to type the start or execute command. Therefore the lights on the tester or on the face of the display will show an almost instantaneous response of pass or fail. If the logic is correct, of course, it is going to pass. One of the tests that could be done now, if you are uncertain about the stability of the system, would be to take the vectors that were previously generated, go in and modify them, and then falsely generate failures. This is a similar procedure that was used in fault generation, except it is the vectors that change, not the faults in the logic.

This ensures that the test program is accurately catching failures. For instance, in the data patterns shown in Figure 9-9, if we changed an output 1 to an output 0, the tester should catch that change and cause a failure. It is very important to do that so there is no false sense of security. This prevents wasting time going through large blocks of logic thinking that everything is actually being tested. One can accidentally set up the machine to ignore failures and make everything pass regardless of whether or not the device is functioning correctly. In addition to that, many of the machines have the ability to invert data

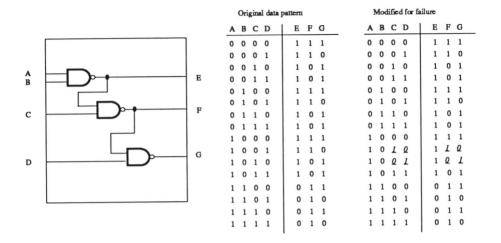

		Original data pattern					
A	B	C	D	E	F	G
---	---	---	---	---	---	---
0	0	0	0	1	1	1
0	0	0	1	1	1	0
0	0	1	0	1	0	1
0	0	1	1	1	0	1
0	1	0	0	1	1	1
0	1	0	1	1	1	0
0	1	1	0	1	0	1
0	1	1	1	1	0	1
1	0	0	0	1	1	1
1	0	0	1	1	1	0
1	0	1	0	1	0	1
1	0	1	1	1	0	1
1	1	0	0	0	1	1
1	1	0	1	0	1	0
1	1	1	0	0	1	1
1	1	1	1	0	1	0

Modified for failure

A	B	C	D	E	F	G
0	0	0	0	1	1	1
0	0	0	1	1	1	0
0	0	1	0	1	0	1
0	0	1	1	1	0	1
0	1	0	0	1	1	1
0	1	0	1	1	1	0
0	1	1	0	1	0	1
0	1	1	1	1	0	1
1	0	0	0	1	1	1
1	0	$\mathcal{1}$	$\mathcal{0}$	1	$\mathcal{1}$	$\mathcal{0}$
1	0	$\mathcal{0}$	$\mathcal{1}$	1	$\mathcal{0}$	$\mathcal{1}$
1	0	1	1	1	0	1
1	1	0	0	0	1	1
1	1	0	1	0	1	0
1	1	1	0	0	1	1
1	1	1	1	0	1	0

Figure 9-9. Introducing a false failure.

patterns both in and out. This was described in the format section of Chapter 2. It would be worthwhile at least doing a simple check with an oscilloscope to verify that the ones going in are truly ones and that the zeros going in are truly zeros; the same is true for outputs.

If the device is totally nonfunctional, this is the time some vendors will do microprobing. This is not an easy task, and requires considerable skill. Check with your ASIC vendor if you need help.

Sequential logic test

Testing logic blocks that contain sequential elements is now the next step in the process of debugging the part. Figure 9-10 shows block B with logic containing sequential elements, timing diagrams, and the vectors required to test the block. The test patterns need to initialize the device before any testing is done. *Initialization* is simply to clock in patterns to ensure that all the states of the device are known before trying to make any measurements. To execute this portion of the test, use the very procedure followed in the two previous examples.

Once again, an opens-and-shorts test ensures that the machine is hooked up correctly. Then follow the routine for the application of voltages to input, outputs, and power connections. Next, a routine for the selection of the multiplexors

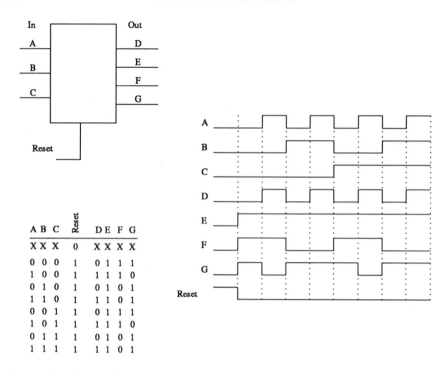

A B C	Reset	D E F G
X X X	0	X X X X
0 0 0	1	0 1 1 1
1 0 0	1	1 1 1 0
0 1 0	1	0 1 0 1
1 1 0	1	1 1 0 1
0 0 1	1	0 1 1 1
1 0 1	1	1 1 1 0
0 1 1	1	0 1 0 1
1 1 1	1	1 1 0 1

Figure 9-10. Blocks with sequential elements.

for the test mode could be used; then start a pattern to clock the proper number of data patterns to ensure that the device is in a known state. During this time, all the outputs need to be ignored because if the part is in an unknown state, false errors will show up at the output.

If the device contained a reset circuit that allowed every sequential element in the device to be reset prior to the forcing of data patterns, life would be a lot simpler. Unfortunately, with pin count being constrained that may be necessary to do in the patterns. One option would be to use one of the test modes of the test mode multiplexor as a reset circuit. In this case, clock in a state which allows the reset of all sequential elements of the part, then go back into another state which selects the block that is about to be tested, and then start forcing data patterns to the part and comparing data. This minimizes the need for long routines to clock the device. These routines to set up and clock the part into known states can become fairly complex. Figure 9-11 shows the sequence used for testing sequential blocks using a test mode buried within a device. The output multiplexor is selected, patterns are impressed upon the part, and the device is tested. It is no

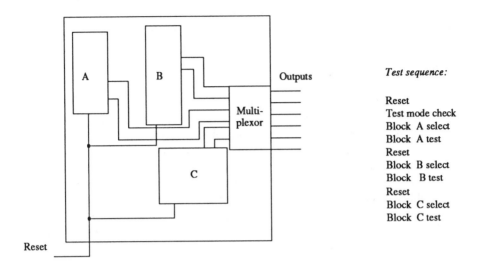

Figure 9-11. Sequence for testing sequential blocks.

different than the combinatorial blocks presented earlier, but the timing and vector relationship is more complex.

Notice that this was built on the test done in previous sections and that the program is now larger and more complex. Keep repeating this program building for all blocks of the device. The last step of the functionality check would be to analyze the functionality of any blocks defined by the vendor as a macro, or large cell, of any kind. Most of these cells would have data patterns tested by the vendor, and it may not be worth the effort to test them at this time of functional verification of the prototypes. If you need to look at them, one alternative is a partial function check. This type of verification would be to take a small subset of the capability of the block and ensure that it is functioning correctly in the system. Again, go through the same kinds of routines, set up the part, the voltages, timings, and formats, and do the opens-and-shorts test. Then force data patterns on the part and compare outputs. For large logic blocks that were supplied by the vendor, it would be wise to run only a very small subset of the data patterns. This is assuming the vendor takes responsibility for testing of the block and that verification means only doing a basic check. If the vendor does not take responsibility for testing the large blocks within the device, then a full check is needed. If so, you must go through essentially the same kind of routines to define the functionality of the part. This includes all necessary setup conditions along with testing. To do that, of course, there needs to be an understanding the internal logic down to the gate level of the block. This will verify that the part is performing the proper function and contains no defects.

PROGRAM ASSEMBLY

If you assume each of these test modules was generated individually, and that now all modules for the debugging of the device have been tested, the next task would be to link them all together. This will make sure that they run together without any interaction that will cause them to have failures, false or real. They can be appended, and in most commercial design verification systems this is a very simple command in the editor. If the block is fully stand-alone, with its own voltage and timing set up, everything should pass. If they do not pass, check interaction of the blocks. Is a timing generator and pin set in one block that is not reset in the next? Does the block test start with a reset? Do you get the same results when manually running the patterns? All these questions can help find the cause.

This step will be the first check of the entire contents of the device simulation. All the vectors should run at a nominal voltage, with no timing or voltage marginality. This should prove that the logic design and the test vectors match.

Checking other devices

Once the test pattern is set up, run the data patterns, and if everything functions correctly, start looking at other devices. This is to check that the other parts that were given by the vendor all function the same way. Assuming there are 10 parts, test all 10 of them. Be careful to monitor the results and perhaps serialize and log the units as to which ones pass and which fail. Also, if any of the test fails, log what type of tests they are.

It is wise to note that these are all MOS devices and that they are highly sensitive to electrostatic discharge (ESD). The normal precautions of handling devices should be taken to protect them. Be sure to be physically grounded. That means wearing a wrist strap connected to the ground of the tester. Preferably work on a bench that has either a metal grounded cover or an ESD protective cover.

Although it may be convenient, do not set them up in an office area with a carpet, even though some of the small verification systems may work well in an office environment. A formica desk is a good insulator, but a walk across a carpeted floor could pick up considerable ESD and perhaps destroy the devices. It is easy to blow up all the devices, and you may end up having nothing that functions. Further, it is easy to destroy the tester—primarily the pin electronics—in the process. The drivers and receivers in the commercial test systems and verification systems are very sensitive. Blown pin electronics cards are the most common type of failure that are seen in these test systems.

Performance verification

After all the functional data patterns have been run, the next step is to test the device for performance. This verifies that the functionality not only happens according to the data patterns that were used for simulation, but that it happens at the right time and voltage. These patterns did not verify that the parametric performance of the device is acceptable for the system use. Input testing is a relatively simple procedure. Assuming that all these patterns are now linked together, the first step would be to use the program setup section and change the V_{IL} and V_{IH} levels from 0.0 and 5.0 V to 0.8 and 2.0 V for TTL, respectively. Other levels (CMOS, Schmidt trigger, clocks etc.) should be changed to the specified value for those pins. These numbers should be the numbers that were in the data sheet as guaranteed by the vendor. After running the data patterns, assume that the data patterns fail and that there are 10 different program modules linked together. The device is failing on the fourth pattern and the tester shows an incorrect output. What is probably happening in the part is that one of the V_{IL} or V_{IH} levels as seen on an input pad is not causing that input of the device to recognize the one or zero input level. Thus, you are using the wrong logic level for the logical computation inside the part. At this time, the tester should at least tell which pin is failing and at what vector.

The next step to do is to find out what went wrong. One can do several tests now. Perhaps first go back to the logic to see whether the output at the wrong state can be directly attributable to an input being forced incorrectly. For instance, is it a strictly combinatorial path? Or is it a sequential path that required some clocking sequence? If it has sequential elements, what is being seen is perhaps a failure of something that happened in a previous cycle. This may not be obvious where it happened from the output states of the device.

For the case of strict combinatorial logic, most testers will stop on the failing vector. Then by using an oscilloscope or perhaps the test program debugging routines, one can go through and look at the inputs and outputs of the device. This is to verify that all the ones and zeros are correct. Several things can be happening here, and assuming that the data patterns passed the functionality check correctly, you can now assume that the data patterns are correct.

Failure modes that show up here are related to misprocessing of the device. For instance, if the threshold voltages are set to a value where ones and zeros are not accurately switched between the 0.8- and 2-V levels, there may be an error in the net list on the selection of input buffers used for the device. For instance, a

CMOS buffer was used in place of a TTL buffer. Or is the test system actually testing the part correctly and is it really representing ones and zeros in the proper manner on that pin?

The easiest approach of course, is the last one to verify, which is to use an oscilloscope and voltmeter and measure the voltage. It is worthwhile, even with the sophisticated bench-top equipment that is available today, to use an external piece of equipment. Items such as an oscilloscope should be occasionally used to verify that the tester is actually forcing what is expected to be presented to the device. Assuming that the V_{IL}'s and V_{IH}'s are correct on the pin and that the failure is attributable to only one pin, then the next step would be to go to the schematics and see whether there is any other logic that was implemented within the part that could cause the failure.

Most likely libraries support both CMOS and TTL switch levels, and the switching level for a CMOS device is outside the range of a normal TTL operation. One test that can be done to find the exact failure mode is to relax the parameters again and see if they pass. Do this on a pin-by-pin basis. If they pass at 0 to 5 V and fail at 0.8 and 0.2 V, try an intermediate value and try working with only one parameter at a time such as V_{IH}. Perhaps set V_{IH} at 3.0 V and see whether the part passes or not. This particular process of debugging can become very time consuming. Realistically not many problems come up with inputs, and this should not be emphasized too much.

If the tester has canned routines for shmoo plots, set one up here for V_{IL} and V_{CC}, or for V_{IH} and V_{CC}. This will tell how close the device is to the specification of the ASIC. It will also confirm when the device passes and fails. The goal is to find the cause, either the tester or the device.

Output verification

The verification of outputs is done in a very similar manner. Assuming that all the inputs now function at V_{IL} and V_{IH} levels together, the next step would be to set up the test program to look at the output capability. Again, there are several choices, one of which is to keep the input parameters set at the known input voltage level or the specification level of 0.8 to 2.0 V, assuming the TTL system, or to loosen them back up to 0 and 5 V, respectively. For the debugging of large devices, it is best to eliminate the interaction and ensure that each piece works by itself individually.

In this case, that would mean loosening up the input voltages to 0 and 5 V for inputs, then changing the output voltages' switching point from a 1.5-V level to the 0.4- and 2.2-V level, assuming the normal TTL data sheet specifications. Once the changes have been made and all the modules inside the program are configured correctly, again run the data patterns and see whether the devices passes. If it does pass, that's fine. If it fails, again, go through the same kind of debugging process to find out what is wrong in either the test system, the silicon, or the programs used for debugging the part.

If you get a failure in a simple combinatorial data pattern for an output low level, start debugging the part with all the necessary setup conditions in mind. Again, stop the tester on the point of failure and look at input and output conditions to verify that the device is receiving the appropriate data as generated by the simulator for the data patterns. Looking at outputs is somewhat easier than looking at inputs. This is due to the fact that most testers allow the probing of pins. You may also use an oscilloscope or voltmeter to measure the output voltage on the part. If you are expecting a 0.4-V V_{OL} level, and getting only 0.3 or 0.2 V, the device should pass. This means that the device is driving better than needed for the system specification. If this is the data pattern is that is expected (logic 0 expected, and 0.3 V measured), there is probably something wrong with the setup of the tester, causing a false failure. This could be noise, incorrect strobe, or marginal timing, for instance.

If the device is at 0.5 or 0.6 V and there is no load on the part, then there may be an error in the configuration of the device in the test system or something physically wrong with the manufacturing of the device: Its output drive capability may be bad, or it may have the wrong kind of output buffer. In addition to that, testing outputs can show up many other problems that may not have shown up in the testing of input levels. For instance, it would be worthwhile to look at the ground for the device versus the ground for the tester. Is the 0.4 V that is being measured, 0.4 V relative to the 0 voltage level of the tester or relative to the ground connection of the part? If the part is drawing excessive current (most CMOS devices do not do so, especially during slow-speed static testing), the ground pin could move up a few hundred millivolts, thus eliminating all the margin that was available in the 0.4 volt V_{OL} test. Very easily a device could have a ground shift of 200 mV if there were a condition such as multiple CMOS input pins that are floating or shorted output pins.

Figure 9-12 shows the typical input structure as implemented in transistors. In the case of intermediate voltages, the device does draw current if inputs are allowed to float. In addition to that if there are buses internal to the part, assuming slow clock times and large blocks driving them, they may cause problems. They may be floating because of the slow-speed testing that is taking place, and that may cause excessive power in the part. If it looks as though the device is drawing excessive power and no large block is unconnected or has floating inputs, it may be worthwhile to look at other devices to ensure that it is not in the condition of latchup or that it is not a defective unit.

The failure identification process is identical for circuits that contain sequential elements. At this time it may be worthwhile if the failures are in a sequential logic to modify the patterns and mask the one particular vector that fails. This is to see whether there is simply one state time that is failing or whether it fails from that point in the test pattern on. Failures of the single-vector type that fall

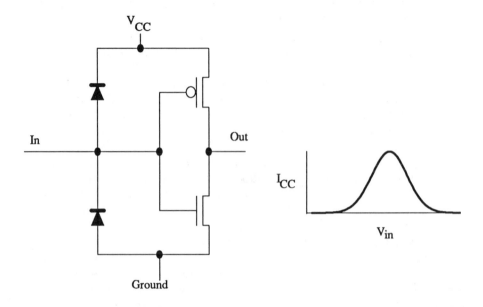

Figure 9-12. Typical input characteristics.

into the category of one failure in one cycle may be related to input conditions on the flip-flops and storage elements in the data path.

Failures that happen in multiple cycles may be related to the logic used to clock the circuit. In some sequential circuits it may shift the data patterns by one clock cycle if done incorrectly. In this case a failure of a one in the first cycle would be a failure of a zero transition in a future cycle based on an incorrect state of a logic element earlier in the logic process. This is often called *cycle skew*. Once all these failures have been identified and understood, go on to a more detailed analysis of the part. The goal here is the same as in input verification: Find and understand all marginalities in the device, and then decide whether to correct them. Again shmoo plots are useful to check margin in the device.

At this time the device is fully checked for all the functionality of the logic and for proper boolean function. In addition, inputs and outputs have been checked individually to make sure that they function correctly to the dc specifications as supplied by the vendor. Next try running a data pattern that was previously avoided, that is, try running both V_{IL}'s, V_{IH}'s and V_{OL}'s, V_{OH}'s set to specification. If they passed individually, they will probably pass together. If they do not, the same debugging procedure should be used, and you should realize that this type of a failure mode is a very uncommon situation. Noise can affect the performance of the device by causing input level buffers to sense the wrong level, output levels to be out of specification, and latches and logic internal to the device to switch states. In addition, noise on output pins of the ASIC will cause readings that look like failures if a window strobe is used at the time of the noise.

TESTER CAPABILITIES AND AC MEASUREMENTS

Before looking at ac measurements on a piece of commercial test equipment, one of the items that needs to be understood is the ability of the machine to drive different formats. The cases discussed so far have been using very simple data patterns that switch essentially at the clock cycle boundary. That means that we were not setting up timing generators to drive inputs at different times within the period nor allowing formatting to take place on an input or tightly strobe output signals of the part. For the purposes of understanding tester capabilities, this section will look at a very simple storage element, a D flip-flop, using the specification parameters of the device to prepare a tester to find out how to measure its ac performance levels. Figure 2-2 in Chapter 2 made another reference to timing generators and specifications.

First, the performance of a flip-flop can be measured without running it at the maximum clock rate. For instance, if a flip-flop runs at 50 or 60 MHz, it can still be tested at 1 or 2 MHz to verify that its functionality is correct. It can check that it meets most of its timing specifications. If you examine the data sheet for a D flip-flop, parameters that are of interest would include setup time, hold time, minimum clock width, and perhaps rise and fall time. A typical logic diagram looks like that shown in Figure 9-13.

In this case, the clock on the C input is represented on the drawing along with the data pattern presented on the D input. We also have the timing parameter of setup time and hold time shown. In addition to that, the propagation delay to the output is shown relative to the rising edge of the clock. This particular flip-flop is a device that stores data on the rising edge of the clock only. Most MOS libraries include flip-flops such as J-K's, D's, R-S's, and simple latches. They may also include shift register elements used for scan testing. Assuming the need to test this D flip-flop on the tester, again, you need the data patterns, the ones and the zeros, and the clocks necessary to run it.

For the purposes of doing a setup time measurement, go through the sequence shown in Figure 9-14. This would force varying levels of the data patterns to be presented at different setup time to ensure that the device captures data correctly, assuming that the data sheet for this example requires the device meet a set-up time of 2 ns and a hold time of 3 ns. The tester in this example has a minimum pulse width of 10 ns. Notice that it would be impossible to test that parameter in one pass on the tester.

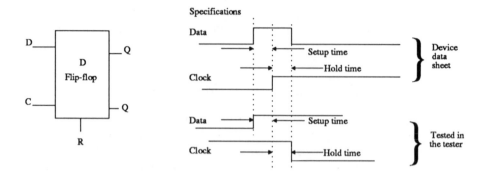

Figure 9-13. Setup and hold characteristics.

The minimum pulse width needed for the device is significantly smaller than the tester can create, and so the tester would need to run through two different data patterns in order to verify that the device is functioning correctly. To verify setup time, go through a procedure as shown in Figure 9-15. Clock a zero level into the part, and then at a known time of 5, 4, 3, 2, 1 ns before the clock changes the input from a zero to a one, clock the device and then strobe the output sometime later.

In the previous sections, there was a discussion about separating the parameters for input and output dc levels. This particular case will do the same for an ac test. This will check the setup time by using a loose strobe and tight clock data. Then check propagation delay using a tight strobe and a loose setup time. Finally check complete functionality with the appropriate setup and strobe time. Next, run the same data pattern looking at hold time. As discussed previously, the minimum pulse width on the D input is significantly lower than the minimum pulse width of most testers. Therefore, the input will be stable for a long time before changing the clock, and later, at the appropriate time, change the input and measure the output. In very high performance silicon libraries, these numbers are in the 1 ns or less range. In many of the commercial testers that are available for bench-top verification, they may or may not be accurate enough to measure less than a nanosecond. There is a need to understand the system specifications of the machine, including pin-to-pin timing capability before attempting to make those kinds of measurements. The type of commercial test equipment used by large manufacturers for these kinds of device measurements may cost millions of dollars.

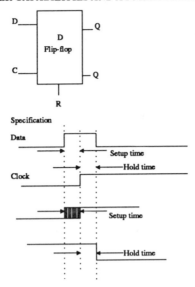

Figure 9-14. Timing searches for ac performance.

A small verification bench-top machine may cost in the hundred-thousand-dollar range. An accuracy specification of 1 ns does not mean that a 1 ns pin-to-pin skew can be measured easily. One of the big differences is the accuracy and repeatability of the system in order to make the measurements.

Now that this simple logic element has been tested and understood, go back to the device and start measuring functionality of the part with respect to certain ac parameters. Figure 9-16 shows the simple logic diagram and the simple logic in block A. One particular path that would be worthy of measurement would be the propagation delayed through the AND-OR logic that was checked in the original block test with the original functional patterns. This figure shows the performance measurement of a block of combinatorial logic. The critical path is identified, as discussed in Chapter 7, and the test sequence necessary to find the actual performance of the part is also shown. In this case the input is moved and

Procedure

Set strobe setup = 5 ns

If pass, Strobe+Strobe-1ns, if not, set fail = 1

5 4 3 2 1

Figure 9-15. Edge search routines.

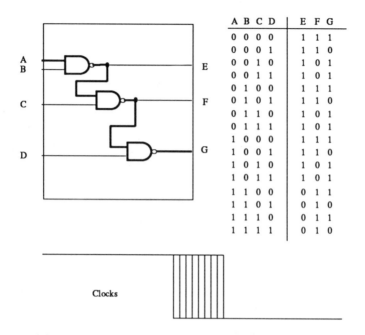

A B C D	E F G
0 0 0 0	1 1 1
0 0 0 1	1 1 0
0 0 1 0	1 0 1
0 0 1 1	1 0 1
0 1 0 0	1 1 1
0 1 0 1	1 1 0
0 1 1 0	1 0 1
0 1 1 1	1 0 1
1 0 0 0	1 1 1
1 0 0 1	1 1 0
1 0 1 0	1 0 1
1 0 1 1	1 0 1
1 1 0 0	0 1 1
1 1 0 1	0 1 0
1 1 1 0	0 1 1
1 1 1 1	0 1 0

Figure 9-16. Performance of a combinatorial block.

the strobe is stable. For combinatorial logic with no sequential elements, moving the strobe or the clock for the data is acceptable.

At this time, to minimize confusion, it is wise to set V_{IL}'s, V_{IH}'s, V_{OL}'s and V_{OH}'s loose. Then run the same data pattern as used in the previous example at the beginning of the chapter, perhaps at 1 MHz again but with the time requirements from input to strobe set at the propagation delay necessary for the logic elements as shown in the simulation of the device. If the simulator showed that this particular path was 13 ns, set the timing delays for 13 ns and see if it works. Shmoo the parameter back and forth to find the margin to specification.

Using the same kinds of technique, one can go on to the sequential logic and the more complex logic blocks within the device, verifying the ac performance. Use the test modes of the device and patterns in order to check functionality. The last step then is to combine input and output and timing parameters all in one case. It should be noted that manufacturers do not guarantee full functionality to all data sheet parameters over the entire process range for devices driven to exact specification levels for inputs and outputs. The levels in the data sheet are dc conditions, and when used for timing purposes, they are not guaranteed. Some manufacturers state that inputs need to be driven to an output level for timing measurements. The timing measurements are made relative to a 50 percent point, for instance, separating the ac performance of the part from the dc specs. This may require the tester to be set to V_{IL} levels at 0.4 and 2.2 V for the purposes of driving signals for ac testing. This is the dc level as presented by the output of the previous stage if driving it correctly. Output sense levels would be set at 0.8 and 2.0 V for ac measurements (or 1.5 V) again, or the input levels as they would be seen in the next stage. AC measurements are then done using those kinds of dc conditions on the inputs and outputs and on the timing parameters as used in the simulation.

Once all these patterns are linked together and run, if the device functions and completely passes all the data patterns presented to the part and all the ac/dc parameters as used by the part, be confident that the device is functioning according to the simulation.

DC parametric testing

The last type of test that has not been checked and may or may not be easily done on design verification machines is the test for the ability of the outputs to drive current specification such as I_{OL} and I_{OH}. This can be done with a parametric measurement unit or by holding resistors on output pads with known drive capability, or by means of *dynamic loads*, if the machine has them. Any of these can actually be incorporated into the test pattern and test the device.

A full functional test of the device at speed with input levels set to specifications, output levels at specification, and timing parameters relative to the simulation with output loads connected to the device is a very thorough test of the performance of the silicon. This will tell whether there are any potential failure modes in the device. Be confident, if it passes this, that the simulation and silicon match 100 percent.

I_{CC} testing

The testing of the I_{CC}, or current consumption, of the device falls into two categories. The first is a static I_{CC} test. Typically this is a test where inputs are held to a stable state and I_{CC} is measured. The measurement can be done by the PMU (most accurate method) or by the device power supplies (least accurate method). Power supplies usually have a sense or trip level that can be used to detect overcurrent values. Using the power supply would require powering up the device and checking the trip in the flow of the test program. These trips most often have no ability to read the actual I_{CC} number but present a pass or fail signal only to the tester. Some of the newer machines have accurate current measurement capability on the device power supplies.

Using the PMU for I_{CC} measurement is far more accurate. The value determined by the PMU can be read directly into the test program as a variable. This means that distributions of I_{CC}, for instance, can be constructed that give a good idea of the typical value. This may be of more value than a strict pass/fail signal. In addition, the accuracy and resolution of the PMU for measurement of current is significantly better than the trip level on the power supplies. This accuracy is especially important in the quiescent test where values of a few microamps may need to be measured.

High values of I_{CC} can be attributed to many factors. The most common are floating nodes (this is either internal to the part or device input pins), dynamic logic that needs clocks running to keep in a low current mode, or dirty connection on the load board or socket assembly. If the value varies dramatically from one reading to the next, it is probably one of the two former causes and not load board leakage.

Dynamic I_{CC} testing

The dynamic value of I_{CC} can also be tested on the part. In this case, the device is cycled and a capacitor on the V_{CC} line is used to average the current spikes. The PMU or the device power supply can be used for this test. It is important to note that dynamic I_{CC} is related to the patterns used. The more nodes internal to the device that switch, the higher the value of dynamic I_{CC}. In addition, the frequency of operation is also related to dynamic I_{CC}. Make sure the patterns and frequency are similar to what will be seen in system operation.

Leakage testing

Measuring leakage is another test that could be added to the test program. Leakage tests are good indicators of misprocessed material and devices that were damaged by electrostatic discharge (ESD). Most ESD damages happens on the inputs and outputs of the device, therefore the current on those pins is substantially higher than normal. A good input will draw less that 1 μA whether it is an input or I/O. This assumes the I/O is in the input mode during the leakage test. A pin that draws 50 μA or greater usually is blown, unless it has a pull-up or pull-down resistor in the pad circuitry.

Testing for input leakage is very simple on devices with nonmultiplexed inputs. No patterns are needed and no special setup is required to test the device. Connect the PMU to the pin where the measurement is desired, and read the value. This returns a value that can again be used for statistical purposes. The range of the PMU should be set to 10 μA if the PMU is not autoranging. A clamp should always be set to prevent damage to the device.

To test for leakage on I/O pins, the device must first be clocked into a state where the pin is in the input mode. Examine the data patterns that were used in simulation and find a suitable vector. Set the stop of the data pattern at this point, and set the tester so that it will not power off the device after the test has stopped. Be cautious that the logic is true and stable at the end of the cycle and that it will stay that way for a long time. Once the test pauses, the program can disconnect the driver and connect the PMU to the same pin. A pause is needed for the PMU to stabilize, and then a measurement can be taken. This is also variable and not a pass/fail signal. This process should be repeated for all input and I/O pins of the device until all of them are tested.

Latch-up testing

Testing for latch-up may or may not need to be done. In many cases the vendor can supply data for the input and output pads and for their latch-up protection characteristics. If testing is done, current is forced into the pad while the device is powered on and I_{CC} is measured. The current causes the clamp diodes to forward-bias. Figure 9-17 shows the cross section of CMOS device and the formation of the silicon controlled rectifier (SCR). Most libraries have sufficient protection that this is not an issue.

Although the figure shown is for an n-well CMOS process, the exact same case holds true for p-well CMOS processes. The formation of the parasitic SCR can be avoided by proper spacing and by placing guard rings around well boundaries to prevent the transistor combination from forming a beta product (n-beta * p-beta) greater than 1. This is the condition needed for the SCR to latch up.

Process variation and characterization

The final parameter to look at is the nature of the characteristics of the device over the process ranges. As noted in Chapter 3, MOS device characteristics dramatically change with voltage, temperature, and processing parameters. More than likely the manufacturer gave prototypes that were processed close to typical processing. That means the middle of the distribution or as close to it as the manufacturer could get, depending on how tightly controlled the process was. Up until now all the work was probably done at room temperature at nominal V_{CC}. Now is the time to verify V_{CC} and temperature performance. Simply look at the device at $0^{\circ}C$ and $70^{\circ}C$ for a commercial specification. Use a thermal probe to hold the device at high temperature and measure the performance of the part at high and low values of V_{CC}. Make sure the probe is accurate. Cold temperature is significantly harder to check because the dewpoint and moisture condensation can affect the readings and the performance of the device.

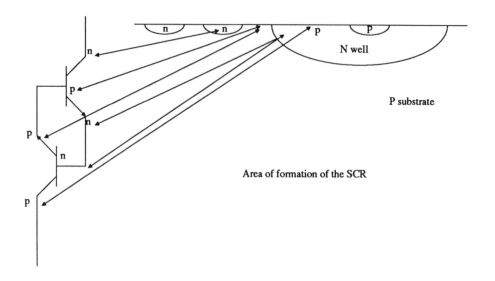

Figure 9-17. SCR formation in CMOS

It is difficult and very expensive to have vendors supply what is normally called *characterization runs*; this is where the process is forced out to the bounds or the edges of the acceptable manufacturing window. The cost of generating those kinds of devices is very expensive, and it should not be done unless absolutely needed.

CHARACTERIZATION

Characterization of silicon has one of two possible representations of the term. As used in this book, silicon characterization comes from a manufacturing background where it is the measurement of the performance of the device with slow and fast processing over the temperature and voltage parameters used for the system design. This kind of characterization is sometimes referred to as silicon characterization or physical characterization of the device. It is usually done with package units at temperature measured on the test system by characterization and shmoo plots.

Another definition of characterization is the software simulation for the silicon performance of the cell or chip. Many vendors and many engineering people interchange the two terms, and so depending on who is talking to whom and on what kind of data is being reviewed, characterization data may be either one. It is important to understand which of the two is being reviewed and understand the impact to the device.

Library characterization

As a final note it would be worthwhile to talk about library characterization as done by vendors. Most vendors will take their library and by using test chips measure the performance of the library elements versus the simulations models to ensure that they are correct. If more confidence is needed that the library functions correctly in your system, ask for characterization data of the library. The vendor should be able to show data that gives slow and fast processing over multiple different temperature and voltage ranges comparing the simulator data with the actual performance of the cells in silicon.

There are many different methods of generating characterization data, and vendors may use customer-designed devices with critical paths that can be measured or they may do specific test chips that check individual library elements. In any case, they should be able to show that the simulation values as used by the simulator at the slowest possible processing, low voltage, and high temperature is outside of the range of the slowest acceptable fabrication material.

Fabrication material is accepted based on E-test parameters. If there is data that shows that, and if simulations were done correctly and the device functions properly in the system, the final step would be to sign off approval of the prototypes and allow for volume manufacturing. Nothing more can be done to verify that the silicon functions correctly and that the design was done correctly. At this time your confidence in volume manufacturing of the device should be very high. After final prototype approval, the production silicon received should be identical or well within the range of what you have simulated and have received so far.

Wrapping up

Assuming the device passes all the tests so far, including temperature and voltage parameters, you are in good shape. This is probably sufficient to verify that the device is completely functional at least as far as the simulations that were done for the logic that was implemented. The next step, of course, is to plug the device into the system, if that was not already done, and verify functionality in the system environment. If that was done previously and the device functions correctly and all the analysis of the data patterns and timings are correct in the device, be very confident that the device will be manufacturable. The device should function correctly over the temperature and voltage variations of the system.

Application functionality

So far work has focused on generation and debugging of the design relative to the tester and the environment where the ASIC device will be exercised and observed. The final test—and it is the most important test—is system functionality, which should be completed in addition to any testing activity related to patterns used for generation of the device. It is the intended application of the system that is the most important for the functionality of the ASIC. There are cases when test programs cannot differentiate between a good and a bad device, and yet the part functions perfectly well in the system. Likewise, some devices completely pass test programs for all levels of detail and all patterns and yet contain basic logic errors that prevent them from functioning in the system. Analyzing and debugging these kinds of failures can be quite a challenge.

It is also possible to have identical devices, some of which pass in the system and some of which do not. In Chapters 3 and 7 there was considerable discussion about the distribution of ac parameters for ASIC devices manufactured on a typical MOS process. These distributions lend themselves to different performance characteristics of the device. And if a sample of devices is delivered, it is possible that some of them could function and some would not. Being able to analyze why they function and why they do not function in the system and in the tester becomes a pivotal point in the understanding of the design process and making the necessary corrections.

If in the process of debugging the device some devices function correctly in the system and others do not, prior to consultation and redesign with the vendor it is best to understand why they do not pass in the system and why they do not pass on the tester. Using the same techniques described in this chapter, isolate the failing modes, the logic, and test patterns associated with that block. If the test program can be modified to the point where devices that pass and fail on the intended application also pass and fail in the test system, progress has been made. It is then easy to go back in using optimization techniques for speed or logic changes to enhance the functionality of the part to ensure that the next iteration of the design is 100 percent functional.

In Chapter 11 will present a discussion of yield versus cost from the semiconductor perspective, but being able to understand the failure modes is the first step in instituting the change that allows maximum possible yield. For the purposes of ASIC device, it is very important to have them yield extremely high. It is quite difficult for a vendor to go in and modify test programs without a thorough understanding of the part.

SUMMARY

This chapter covered the use of the tester and test modes to debug the device. There were two areas that were attacked, although they may not be obvious. First this chapter verified the functionality of the device by use of the proper test modes and the tester. Secondly, and a lot less obvious, is the fact that the test modes can help identify design errors by giving access to the internal portions of the ASIC.

If the errors in the design are uncovered accurately and are completely understood, the next revision of the silicon will be 100 percent functional. If there was not sufficient access for test and observation of the internal logic, analysis of the root cause of the logic problem may turn into a guess. In this case it may be luck to find out all the logic problems that should be uncovered. This is another reason for good testing.

10

VOLUME PRODUCTION

At this point the device has been approved, accepted, and transferred to the vendor for volume production. This is the point when many of the steps that were recommended in the previous chapters including package consideration, production transfer, and electrical marginalities, will impact the design. This impact will be seen in the levels of service that will be received from the vendor. The first consideration is package availability.

PACKAGE VARIATIONS

Some vendors ship prototypes in a ceramic equivalent package of a plastic product if plastic is what was chosen for production. There are slight variations in packages from ceramic to plastic equivalence in the electrical performance of the package. If the package is one of the surface-mount packages, such as quad flat packs, there may be some problems with the package related to handling. Package selection has a sizable impact on cost and PC board density. Table 10-1 shows a variety of different packages.

Fine pitch packages pose problems in production relative to handling for testing and placing the devices on the printed circuit board. If using conventional manufacturing techniques for the small user, stay closer to the through-hole type of packages and the larger lead spacing packages. Packages such as EIAJs and PQFPs that have very fine pitch interconnection points require automated equipment for the final placement of the package.

Any manual handling of the package has the tendency to degrade the lead coplanarity and lead spacing characteristics of the device. Figure 10-1 shows a PQFP and its coplanarity specs. If you compare that to a dual in-line package, notice that the manufacturer must hit significantly tighter numbers. In addition,

Table 10-1. Typical package data.

AMKOR/ANAM ESTIMATIONS ON MIN/MAX DIE SIZES
PER PACKAGE TYPE

PACKAGE TYPE	LEAD COUNT	MINIMUM DIE SIZE (MILS)	MAXIMUM DIE SIZE (MILS)	COMMENTS
PDIP	16	030 X 030	145 X 500	
(300)	24	040 X 040	150 X 500	
PDIP	24	040 X 040	390 X 600	Note 2
(600)	40	070 X 070	390 X 650	Note 2
	48	100 X 100	390 X 650	Note 2
PLCC	44	070 X 070	400 X 400	
	68	105 X 105	700 X 700	Note 2
	84	140 X 140	900 X 900	Note 2
PQFP	84	115 X 115	400 X 400	Note 1
(Bumpered)	100	150 X 150	500 X 500	
	132	225 X 225	700 X 700	Note 2
	164	260 X 260	900 X 900	Notes 1 and 2
	196	320 X 320	1000 X 1000	Notes 1 and 2
QFP	44	050 X 050	200 X 200	
(EIAJ)	64	080 X 080	350 X 350	
	100	160 X 160	350 X 580	Note 2
	160	280 X 280	850 X 850	Note 2

Notes: 1. These packages are currently designed but not tooled at Amkor/Anam facilities.

2. Mechanical stress is a major concern on large die sizes. Amkor/Anam has no reliability data on die sizes larger than 500 mils.

Package estimates courtsey of Glen Koscal, Amkor Electroincs, and John Heacox, Amkor Electronics.

packages such as dual in-line packages have relatively thick lead frames with widely spaced leads. These can easily be modified by a person with a pair of tweezers. PQFPs are fine enough that it is difficult to modify the packages to make them meet spec.

If buying a PQFP package and not handling it automatically, you have several alternatives, one of which is to subcontract assembly steps to an outside vendor who can mount the devices on the PC board. If the PC board contains no other such devices as a surface-mount high-density package, perhaps you should avoid buying this kind of a package.

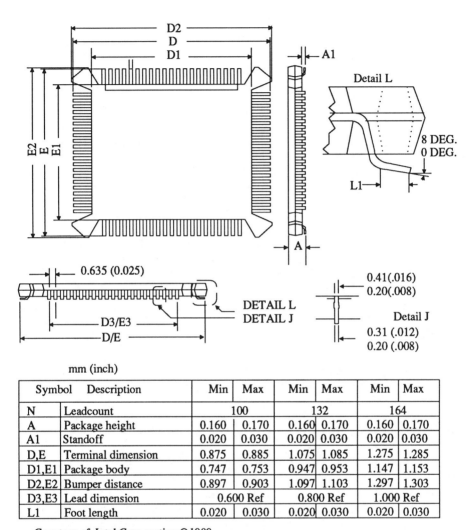

mm (inch)

Symbol	Description	Min	Max	Min	Max	Min	Max
N	Leadcount	100		132		164	
A	Package height	0.160	0.170	0.160	0.170	0.160	0.170
A1	Standoff	0.020	0.030	0.020	0.030	0.020	0.030
D,E	Terminal dimension	0.875	0.885	1.075	1.085	1.275	1.285
D1,E1	Package body	0.747	0.753	0.947	0.953	1.147	1.153
D2,E2	Bumper distance	0.897	0.903	1.097	1.103	1.297	1.303
D3,E3	Lead dimension	0.600 Ref		0.800 Ref		1.000 Ref	
L1	Foot length	0.020	0.030	0.020	0.030	0.020	0.030

Courtesy of Intel Corporation ©1989.

Figure 10-1. PQFP mechanical specifications.

If you are planning to handle the devices in-house, one item worthy of investigation at this point is the cost of handling. Figure 10-2 shows the tradeoffs of physical handling of the device and the yield associated with handling, versus the cost of the device. Notice that, with an expensive device, a degradation in the yield of the product effectively increases the overall cost.

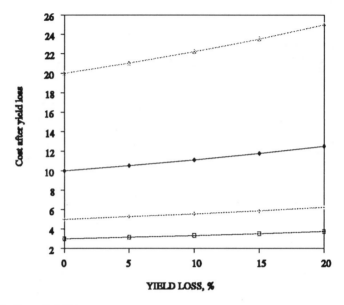

Figure 10-2. Cost of yield losses.

Semiconductor companies invest a lot of money in package handling to ensure that the yields are as high and as repeatable as possible.

PRODUCTION INTRODUCTION AND TRANSFER

After receiving formal written approval of the prototypes, the vendor will transfer the product into production. Depending on the type of process and the vendor's capability, the formal transfer procedure will entail several activities related to specifications, fabrication, and documentation. The costs associated with these transfers either may be included in the original nonrecurring engineering (NRE) charges for the device that was paid for as samples or may be part of the production manufacturing cost amortized over the volume of the device that is expected to be purchased.

Fabrication

The first items that the vendor will evaluate are the purchase order, the contract, and the manufacturing tradeoffs for the mask set. The originally purchased mask

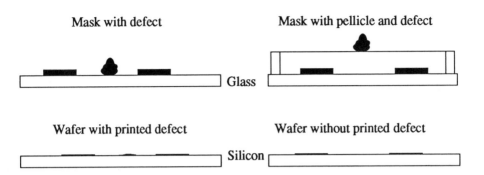

Mask with defect   Mask with pellicle and defect

Glass

Wafer with printed defect   Wafer without printed defect

Silicon

Figure 10-3. Defects on plates.

set may have been of prototype quality. There may be backup sets required to ensure a viable long-term manufacturing flow.

Ordering a second mask set or replacing the mask set with a high-quality plate set is one alternative. In addition, the vendor may be buying what is called a *pellicle set.* Pellicles are basically a clear material placed over the chrome side of the glass to take particles that may have landed on the plate out of the focal plane of the lithography equipment. Figure 10-3 shows a cross section of a plate focused on a wafer, revealing the effect of a pellicle and the ability of the lithography equipment to repeat that defect on the wafer. Note that the depth of field of the lithography equipment is in the micrometer range.

Mask sets are inspected in detail for defects and critical dimensions. This inspection uses very accurate measuring machines, and so the probability of a defect showing up on a production plate set that did not exist on a prototype plate set is very small. A mask could have a defect that impacted the device on the prototype plate set that made it work correctly. That defect may not exist on the product plate set and, therefore, could produce different results in the silicon. But this is probably not worth worrying about.

The next item that the vendor will look at is the wafer size. Is the prototype manufacturing line the same line as the production line? Some vendors have the capability or requirement of running prototypes on one wafer diameter and of producing production material on another wafer diameter. In addition, there may be many different designs on one wafer. This would not be acceptable in manufacturing for high-volume production of one individual design. If prototypes were generated by combining multiple designs on one wafer, the mask would need to be completely regenerated. Even if for volume production all the devices that were prototyped together went into production together, it would still need a change. In that case, yields from device to device would vary

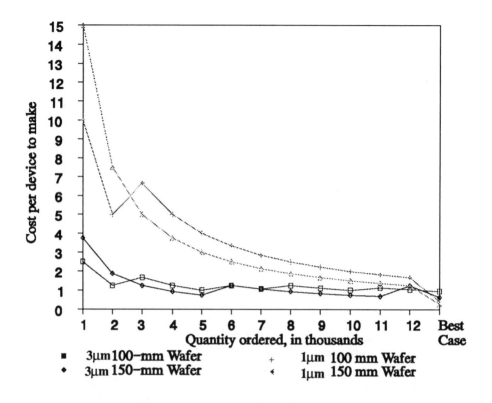

Figure 10-4. Cost of 150- and 100-mm wafers versus volume.

slightly, and it would be in your best interest to go ahead and regenerate the mask set anyway.

If wafer capability is converted from 100- to 125- or 150-mm, performance of the silicon should not change. If the vendor is doing such a conversion, it would be worthwhile to ensure that the characterization of a library discussed in the previous chapters was done completely. Check that wafer diameters are correct and that the prototypes after manufacture meet the same speed and power requirements as production material. The physical area of a 100- versus a 150-mm wafer is about a 2.5 difference in wafer area. For devices this means a comparable difference in the number of die available to be manufactured on the wafer. The cost of manufacturing a 150-mm wafer over 100-mm wafer is greater, but it is not two-and-a-half times greater.

Therefore, the process costs of manufacturing on a large wafer size would be lower than for a small wafer size on a per-device basis. Unfortunately, that is not as simple as it may seem in that the volume characteristics again affect cost. If

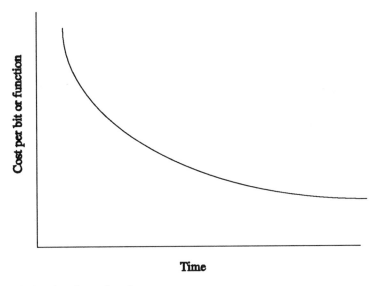

Figure 10-5. Semiconductor learning curve.

buying a certain amount of material based on die size, volume, and wafer diameter, you may be better to go with one size process versus another. Figure 10-4 shows the tradeoff of two different die sizes and the relative manufacturing costs for a 100- and a 150-mm wafer. Notice that the 150-mm wafer is cheaper long-term, but that the 100-mm wafer may be cheaper for the initial introduction. The primary area of concern here is the cost for less than one fabrication lot of devices. Note that in this example the NRE associated with the design is not included.

Semiconductor learning curve

The use of a learning curve is a phenomenon taking place in the semiconductor industry; it is used to predict cost and yields over time. The learning curve assumes that costs decrease by about 0.85 for every doubling in volume. As the process matures, yields are expected to increase. Going back many years, most devices decreased dramatically in cost. The decrease came about because the defect density and the ability to control the process improved with better and better equipment over time and as companies manufactured more and more devices. The defect density and the manufacturing quality improved such that the overall cost went down. Figure 10-5 shows the impact of yield and costs for a typical manufactured device relative to the number of years in manufacturing. This is the semiconductor learning curve. This learning curve is what is primarily responsible for the dramatic decrease in semiconductor device costs over the

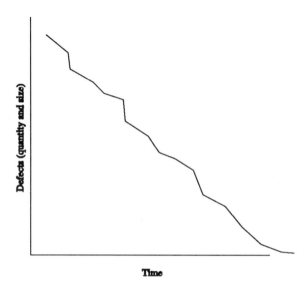

Figure 10-6. Defect density over time.

past few decades. The continual downward trend in cost is sometimes offset by an increase for the short-term. This is the wafer diameter conversion.

The next major conversion in the semiconductor industry will be the conversion from 150- to 200-mm wafers. Although some manufacturers have already started working with the 200-mm size, they are producing such a small percentage of the total volume that it will probably be hard to find this size when talking to ASIC vendors. In some semiconductor companies, the learning curve for ASIC devices is driven off standard products. It would be worthwhile to ask the vendor what is the main process vehicle for defect density reduction. How do they find defects, and what are the processes used for improving general yields? Using an ASIC device as a yield defect density monitor is a very difficult way to achieve yield improvement.

Defects are found by probing and microscopically analyzing individual devices that fail. The engineers in the semiconductor company will do a detailed failure analysis of the bad devices off the wafer to try to find what is wrong with them in order to improve the overall yield. Most vendors use a device that is easily understood and can be very easily mapped, such as a static or dynamic random access memory. This memory is used to drive the semiconductor learning curve. If they do not have a device like that in production, then it is worthwhile to ask how they go about getting the die and wafer yields as high as possible.

This drive for defect density impacts many items in the semiconductor process. Yields are related not only to defect density, but also to reliability and, to some extent, to quality. The quality impact on defect density also is affected by what kind of test coverage was generated for the device. Was the pattern fault graded for the device? Is the part being fully exercised? On a very stable mature process with very high yields, defects are almost nonexistent, and therefore the impact of defects on reliability and quality are extremely small.

A poorly tested device or device tested with minimal fault coverage will probably have good yields or at least have acceptable results on a process with very low defect density. That same device would, on a process with high defect density, yield better than it should, and probably not show many defects in the production material. Defect density is related to process types. Figure 10-6 shows defect density versus year of introduction for processes. More mature processes are lower on the curve than newer technologies.

Production testing

The next step in the manufacturing process is for the vendor to transfer production test capability from the engineering, or prototype test area, to the production test area. That entails several items, one of which is hardware-related. This would be the transfer of the documentation and maintenance capability for probe cards and automated handling machines, including pin connections and physical hookup equipment for the device to be manufactured. Depending on the pin count of the device, this may be a relatively simple or it may be a very complex task. The next step is the finalization and transfer of the test program. The test program that is used to debug the silicon in the previous example is not the best one to run in production. Chapters 4 through 7 discussed the tradeoff of fault grading versus volume production of a device. They also emphasized the need for two different types of programs. The patterns that were generated during the simulation process are what would be run in manufacturing. Once again, it's worthwhile to look at volume versus test coverage.

As mentioned in the previous section, yield has a big impact along with defect density on the overall manufacturing cost. Test programs that were developed for debugging may not cover all faults necessary for a high-quality, low-defect production test program. If there were two programs generated, as suggested in Chapter 7, it would be worthwhile to ensure that the proper one is the one transferred to production for testing purposes. The production test program that functionally checks all devices and is the output of the fault simulation that was done in the previous sections is the ideal one for production test capability.

If there is access to both vector sets and they were both used during the verification cycle as in Chapter 9, do a correlation. There should not have been any cases of devices passing one set of vectors and failing the other. There is, of course, the possibility of appending the two sets together, but that brings back

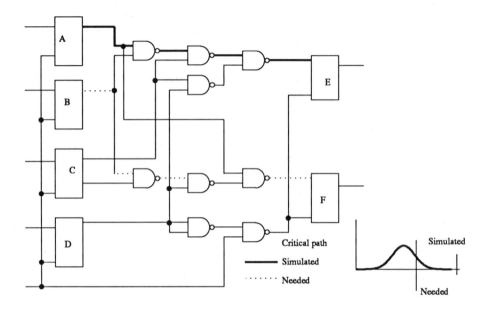

Figure 10-7. Critical path selection problems.

the tradeoff of production test time or vector length versus cost. It should be possible to make one set of vectors completely cover all the aspects of the device to ensure full functionality and good testability in the device without having to append the two programs together.

Returning failures and error recovery

After the device is transferred into production and production quantity of units are received, there are there are still potential problems. During the process of system assembly, debugging and shipments, there may be areas that were not tested even though the fault coverage was relatively high. There may have been mistakes in the logic of the device. Or with a 95 percent fault coverage, it could be that the few percent that were skipped were the ones which turned out to be critical. Today with good production test programs, vendors are talking about defects per million (DPM) devices shipped. Therefore, if buying, for instance, 10,000 devices and the vendor is in high volume production with the process running 100 DPM outgoing quality level, then expect to find one defective part in the quantity of units shipped.

Table 10-2. Cost of early or late yield loss.

Wafer cost*	$600	$600
Good die	200	160
Die cost	$3.00	$3.75
Assembly cost	$2.00	$2.00
Test yield	0.76	0.95
Total cost	$6.57	$6.05

*Assumes a 1μm 150-mm wafer; costs courtesy of ICE.

If there are more defects than the acceptable number as quoted by the manufacturer in DPM levels, it would be worthwhile to go back to look at the test set for accuracy. Also review the simulations suite that was used in the evaluation of the device. If it is obvious that portions of the device were skipped in the simulation process or during fault grading, it may be time to go back in and change the vector set. If the changes in the vector set have a relatively small impact on the device yields, you may want to do a revision of the test vector set.

If there is a logic error that causes the device to fail or a testing error that causes the device to yield very poorly, then new vectors are needed. If the new vector set causes yields to drop dramatically, it is probably wise to assume a new design or additional charges. The manufacturer assumes a certain yield in the manufacturing of the device. If the changes to the production test capability are done in order to increase the customer's yields, the cost associated with the device would go up. More than likely sections of the contract originally relating to purchase would be void. This could wind up being one of the horror stories in which the product cost as quoted by the vendor is greatly increased due to a seemingly minor change in testing.

Figure 10-7 shows the critical path for one of the earlier circuits. Also shown is the actual needed timing for the system. Notice that changing the timing on the critical path test to the number needed for the system means that 20 percent of the material needs to be thrown away. This means that the cost goes up by a comparable amount. This error in critical path selection could be related to gate delays or to parasitic conditions in the device. Table 10-2 shows a cost calculation for a device using high and low yields at the wafer and test step. Notice that this is not only throwing away the die, but that the yield loss associated with the package now becomes a major factor. As you can imagine, from this example with 20 percent of the material thrown away at wafer or package level, the effect on cost with expensive packages would be dramatic.

In older, more mature technologies and high pin-count ceramic packages, packaging is the most expensive part of the total cost. If it is necessary to throw these devices away at the test step and not at the wafer level, costs can be phenomenal. Cost relative to the overall unit cost goes up, and so it is very important to ensure that the test generation as done for prototype and simulation purposes, is as accurate as possible.

These high pin-count packages do not lend themselves to any kind of reworking or modifications to the die or physical changing of the die. Therefore, if a sizable amount of material is coming in at a poor yield, assume that the devices are thrown away. They can be thrown away not only at the user facility, but by the semiconductor manufacturer. This will be one of the major driving forces in the consideration of a redesign.

Speed testing in production

Another failure mechanism that the devices may have is the performance of the library relative to system specs and the need for speed testing. Depending on the margin of the design that was implemented, it may or may not require the implementation of speed testing. Testing at speed is a very difficult process in production.

Once test patterns start running greater than about 20 MHz, test system noise considerations become major factors. You need to have the equipment to maintain sufficiently low noise levels to make accurate measurements, which at high speed becomes a fairly difficult task. Most manufacturers of ASIC devices will try to avoid testing at high speed to eliminate noise problems associated with high-speed testing. If you are in a situation when it is needed to have high-speed testing, there may be additional charges for implementation of the tests for the device.

Temperature testing in production

In certain cases, such as in military applications, there is a need to have high-speed testing over the temperature range. This can become a serious problem, and the physical transfer to production may be delayed simply in an effort to get high-speed testings up and running. If the critical path of the device was dominated by failures related to cycle time, there may be no alternative but to have the device tested at high-speed and not at a relaxed rate. Again, relating back to the previous chapters on design, one of the parameters that impacts high-speed testing is I/O switching and the noise that is generated by it. If the package is made with normal bonding and manufacturing techniques, the inductance and resistance of the ground lines become important for noise generation during high-speed testing. It is worthwhile to look at the physical characteristics of the package relative to the speed and system requirements. The vendor should be able to help with this.

Remember that in a system PC board grounding capability is usually better than it is in a test system. Further, the output driver of the adjacent product could be located physically close and impedances well-controlled. The output of typical devices in systems do not need to have the ability to vary levels (V_{IL}, V_{IH}, etc.) such as would be needed in a commercial piece of automatic test equipment.

Typical production test programs

Typical production test programs are generated using the flow described in Chapter 8. The machines may be complex and quite costly. Typically different test programs are written for test, wafer sort, and perhaps final outgoing inspection (QA). These test programs are all derived off the basic simulation that was submitted with the net list. Test parameters are guardbanded for the different types of tests, and the amount of guardbanding is related to the machine chosen, the process, and the temperature stability of the test environment. Table 10-3 shows a typical test program and the guardbands needed for some of the parameters of the test; but this is only an example. The actual list of parameters tested is usually much longer. Notice in the table that parameters are loosened up relative to the performance of the part and its sensitivities, and is not the absolute value of the parameter. For instance, V_{IL} is tightest when it is above the 0.8-V level for normal TTL specification.

SUMMARY

This chapter looked at the time when all the verification and design work paid off. If the device simulation was done with testing in mind and if patterns were generated for the tester, all should be well. None of these items should become problems that would impact the quality and delivery of the device that's being made. However, if only minimal attention was paid to the testing sections earlier in design of the device, then there probably will be problems. During simulation and prior to net list acceptance was the time the problems should have been found. Problems now may cause the program to require a major reset and have deliveries slip out. Remember that if a new design is required, in most cases this is a new throughput time that will require new silicon debugging. Depending on the size and magnitude of the change, the time lost can of course vary. It can

Table 10-3. Typical fully guardbanded test parameters.

TEST PARARMETER	DATA SHEET	TEST VALUE	QA VALUE	LOOSE VALUE
V_{CC} nominal	5.00	5.00	5.00	5.00
V_{CC} minimum	4.50	4.40	4.50	5.00
V_{CC} maximum	5.50	5.60	5.50	5.00
V_{IL}	0.8	0.85	0.8	0.0
V_{IH}	2.0	1.95	2.0	5.0
V_{OL}	0.45	0.40	0.45	1.0
V_{OH}	2.2	2.25	2.2	1.5
I_{CC} dynamic	2.5 mA	2.4 mA	2.5 mA	10 mA
Access time	40 ns	38 ns	40 ns	900 ns
Setup time	8 ns	7 ns	8 ns	25 ns

require a few weeks of design plus the throughput time or it could take months of rework that could push out the program many months to a year. In addition expect to be charged for more NRE. If the device was designed with testability in mind, you should not have this problem now.

11

TOTAL SYSTEM COST

This chapter will cover some of the aspects of the quality, reliability, defects, and other manufacturing concerns related to the testing of ASIC devices. Emphasis will be placed on them impact of these devices on the quality and overall usefulness of the device that has been designed and is now in manufacturing.

OUTGOING QUALITY

Quality, as defined by most semiconductor manufacturers, is a measure of the grade of the outgoing product, usually stated as acceptable quality level (AQL). AQL is usually expressed as a percentage, such as 0.1 percent AQL. This number represents the maximum number of defective devices that will be contained in the lots that the manufacturer ships. Therefore, 0.1 percent AQL means that less than 1 part in 1000 will be defective upon inspection. Defects per million (DPM) is the actual *number* of defective or bad units in the lot. DPM, AQL, and yield are very definitely related. Sampling plans are used by most manufacturers to ensure that sample sizes guarantee the maximum defect level.

A 0.1 percent AQL used in an outgoing inspection of a high-quality low-defect manufacturing process will guarantee well less than 1000 defective parts per million. These defective parts per million fall into all categories and most manufacturers will separate mechanical defects from electrical defects. Electrical defects are related to logic and electrical performance of the device where it fails to meet spec. Such defects are related to parameters that are within the scope of the test program generated by the vectors and shell from the manufacturer to test the device.

If there is a defect in the device that was not caught or a pattern that was not exercised and observed within the device, this may result in a significantly higher defect number. Once again, the effort put into the test program vector generation becomes a major factor in the ability to meet the low defect levels required by most customers. Once shipments have been made for a while—and the end system is shipping—there might be some failure modes that were not covered in the test sequence originally developed for the device. Again, there are a few alternatives.

OMISSIONS OF SIMULATION

If through an error of omission some path was not checked, fault-graded, or simulated in the original test program sequence, some manufacturers will allow a revision of the test program. If it is a portion of the device that is in the test program and should be caught by the manufacturer using the test programs that were supplied during vector generation and final simulation, then return the devices to the vendor and ask why the defects were not caught. If the area was not simulated in the device and there is no ability to measure it, it may be very difficult to ascertain which type of failure mode it is. Moreover, if there is no verification system available and no particular method to analyze the failure, you have a big problem. Most manufacturers, by the basic nature of ASIC devices, do not understand the logic operation internal to the part. So if there is a failure and it is sent back to the manufacturer, they may reject it. If it is relative to a system specification that the device fails, the manufacturer may not be able to analyze and correct the problem. Figure 11-1 shows a typical logic block diagram layout and schematic. It shows a failure at some point in the schematics and a comparable die layout with that failure pinpointed on the die. The ability of a manufacturer to find the error among 20,000 other gates is virtually impossible without a thorough understanding of the part.

SUPPLIER FEEDBACK ABOUT TEST PROBLEMS

If you have the ability to exercise the part with some kind of design verification system, this system can be used to isolate the failure. Isolate the failing node by exercising it with a very brief data pattern. By doing this and by supplying the data pattern to the manufacturer, the ability to analyze the part, and understand what the failure node is, is of extreme value to the manufacturer. In this way the manufacture does not have to understand the logic versus layout of the device. If a failure mode shows up on certain devices and does not show up on others, ship both good and bad units to the manufacturer with a data pattern that will separate the two, and show the failure mode. Of course, the shorter the data pattern and

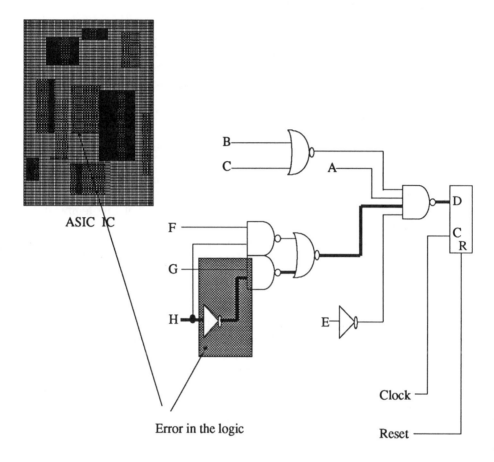

Figure 11-1. Finding logic errors.

the easier it is to understand, the higher the chances of successful implementation. Granted that in a high yielding process this may mean only a few defects per million, but when striving to achieve a high-quality product, this is still a important. This is a very important aspect of device quality improvement.

Most manufacturers have formal procedures for the releasing of test programs to production. If there is a need to change the device, you should be aware that some manufacturers have rules as to how often the user may change programs. It may be worthwhile if there are a few failure modes, or if the failure rate is not very high, to accumulate failures and send them all in at one time. This is better than trying to revise the test program every month or so.

TEMPERATURE TESTING

Most manufacturers will test ASIC devices at a single temperature applying the program in a unit package form. This may be room temperature or high temperature depending on the type of device. If the system is designed with margin, that probably is sufficient to correlate back and guarantee performance over the entire temperature range.

MOS transistors have the characteristic of slowing down rather dramatically as temperature increases. So, the best condition for testing a MOS device for ac performance is at high temperature and low voltage. The best way to test a MOS device for maximum current capability and certain input levels is at low temperature and high voltage. Unfortunately, temperature handling of a part is a fairly complicated aspect, and with the newer generation of packages, testing at high temperature becomes a fairly difficult task. Testing at low temperature, below dew point or below freezing, is even more difficult.

This test capability of below zero or dew point temperature poses the problems of condensation of moisture on the device that will cause false failures. So unless they are required to test the device at temperature via contract, such as for military and extended temperature range applications, the manufacturer probably will not implement cold-temperature testing. One alternative flow is to test it at a hot temperature at one place in the manufacturing cycle, such as during wafer sort, and at room temperature at the final test. Room temperature is only 25°C from the lower extreme of the operating range, and with a slight correlation of dc values, part performance should be sufficient to cover the entire specification range. Alternating current values are usually not a problem as the device typically operates faster at cold temperature as opposed to hot temperature, although race conditions could still cause a failure.

If the defects or system-related problems that are being found are temperature-related, it would be worthwhile to go back to the design verification system which was used and check the device performance compared to the system specifications. It may be that the device is completely meeting the system specifications, and some other component in the system may be a problem due to noise.

If there is marginality that shows up at temperature, there is a chance to change the testing conditions at the manufacturer. This allows for temperature testing at a particular timing and voltage value that would cover the failure mode as discovered in the system application.

Military and extended-temperature range parts

For military and extended-temperature range parts, several aspects change relative to production. First of all, if it is a Joint Army Navy (JAN) device, it will be tested at all operating temperatures, -55°C, +100°C or +125°C, and probably room temperature. This is a requirement and the manufacturer should supply this as a normal means of doing business. The added cost of military devices includes such testing. In addition, it includes burn-in, which will be discussed later. Military libraries are typically relaxed in comparison to the data sheets and specifications for a commercial library. Do not expect the same performance out of the device over the wider temperature and voltage range of a military product.

RELIABILITY CONSIDERATIONS

Reliability is defined as the long-term operation of the device. If quality concerns whether the device functions correctly when it is initially plugged into the system, then reliability zeros in on the failures that show up after an extended period of operation. "Extended" may mean as little as a few hours or as much as 20 years.

Typical manufacturers express reliability as two different numbers. The first is infant mortality, or that failure rate during the first 168 hours of operation while being burned-in at 125°C or during the first 6 months of operation at room temperature. The second number, or long-term failure rate, is defined as the failure rate over an extended period of operation such as 1 to 2 years of typical device operation, or 1000 hours of burn-in at 125° C.

The measurement criteria for infant mortality is usually shown as a percent of failure per hour of operation, such as 0.1 percent per 168 of burn-in at 125°C, or 0.1 percent failure rate in 6 months of normal operation at 55°C.

Long-term reliability is usually defined in terms of failures in time (FIT). Numbers like 100 failures in 10^9 hours of operation are not uncommon. The manufacturer of the ASIC device should be able to supply the information on the infant mortality and long-term reliability of the process and cell library. If the infant mortality number is not sufficient for the system needs, then perhaps you should add burn-in for the part. Once the decision has been made to do burn-in, there are several factors to consider.

BURN-IN

Static and dynamic burn-in are two types that are offered by most manufacturers. Static burn-in applies power to the part at an elevated temperature, but does not exercise the internal nodes of the device. Dynamic burn-in not only powers up the part at temperature, but stimulates the part with data patterns while at temperature. These data patterns are not for testing, but just make certain that a large number of the device's nodes are exercised. Dynamic burn-in is considered to be the best. Dynamic burn-in exercises the devices fully during

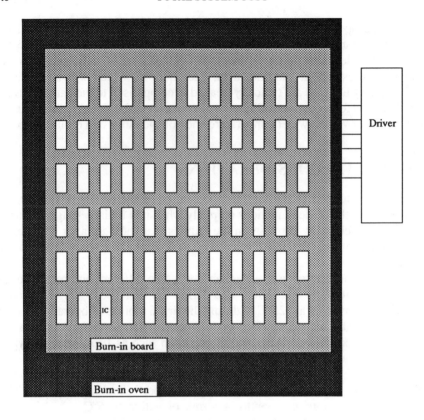

Figure 11-2. Typical burn-in board.

functional burn-in, but that means that a custom burn-in board needs to be designed. The necessary circuitry to actively exercise the device during burn-in also has to be implemented in the burn-in system. Figure 11-2 shows the configuration of a typical burn-in board as used by a manufacturer and some of the electronic components associated with it. Notice that the density of the devices would be related to pin count and package type.

A burn-in board has sockets available for the devices to be exercised. The driver circuit contains a program and input drivers along with V_{CC} supplies to exercise the device. The best possible way to burn-in the part is to run the same data patterns used in the tester for exercising the device. This exercising does not need to monitor outputs of the device.

Burn-in tradeoffs

What happens during burn-in is an attempt to accelerate failures in the part if there should be a weak feature or defect that is sensitive to the extended operation of the device. Items such as weak oxides, narrow metal or silicon lines, small contacts that are resistive, and other failures are usually weeded out. Burn-in may cost a fair amount of money, depending on the type of device that is being burned in. It is more related to the type of package that is running and the pin count of the part. The basic PC board is a multilayer board size of about 10 by 20 inches and, depending on the pin count of the part, may vary from as few as 15 to as many as 200 devices on one single burn-in board (see Figure 11-2).

If the burn-in board costs $2000 to manufacture and it is good for a certain limited number of burn-in cycles, this may be considerable cost per part burned in. A high pin-count part is significantly more expensive to burn-in than a low pin-count part. Therefore, it is very important to make sure that the manufacturing process is reliable without having to go through the necessary steps of burn-in. Burn-in has a direct impact on system costs, and it may be worthwhile, depending on the type of application, to look at the needs of weeding out these kinds of failures.

Once again, looking at the failure rates of the devices over time, a typical manufacturer will represent failures on what is known as a *bathtub curve*. Figure 11-3 shows a typical bathtub curve for a manufacturing process of a CMOS technology. Failures in the first period of time are relatively high, dropping down to a low run rate. The failure rate stays at a low run rate for a relatively long period of time. The x axis is a logarithmic axis, and at this time most manufacturers of highly reliable processes do not know what the long-term failure, or wearout, mechanisms are. The first portion labeled "A" is infant mortality; "B" is the normal operating range of the part; and "C" is the wearout time. For most devices, if designed properly, wearout times are too far away to be measured directly. If a device is designed with sufficient margin, there is really no reason to worry about failure mechanisms in the wearout category. Parameters that can impact wearout, from a integrated circuit design perspective are items like current density through metal lines, total power dissipation, and other factors that would impact the device and its long-term reliability.

If this is a military application or high-power part, one item to pay attention to is the 150°C maximum junction temperature. If the device exceeds that number, it may have a wearout mechanism in the part that affects the reliability of the device. In addition, metal lines that are sized incorrectly for the power dissipation or current draw may cause problems. Most manufacturers that are developing cell libraries will lay out cells with V_{CC} and V_{SS} connections based on a maximum theoretical operating rate and size. Size and clock rate determine I_{CC} in a CMOS device. If the design is pushing the manufacturer's process, meaning

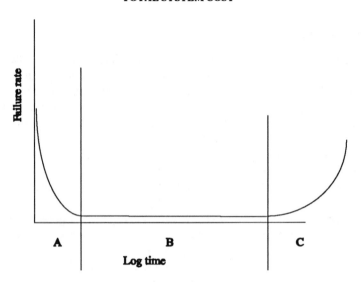

Figure 11-3. Bathtub curve.

that it is running at the very maximum speed in the largest possible die, this may be close to the threshold of maximum allowable power dissipation, or it may be near the maximum current density in the aluminum lines. It would be worthwhile to sit down with the manufacturer and calculate whether the device is getting near the maximum speed and power requirements of the library.

QUALIFICATION

At this point it may also be wise to review the manufacturer's qualification data for the cell library or gate array that was used in this design. "Qualification" consists of a series of stresses that exercise the device to ensure functionality over the entire temperature and process range with reasonable reliability. Typical stresses that are run in a qualification include infant mortality, or burn-in, as was discussed above; long-term life testing, typically going 1000 to 2000 h of high-temperature burn-in; temperature cycling; moisture testing; and bake, or high-temperature storage. These tests are defined as the following:

ELT

The Electrical Life Test (ELT) correlates to extended system operation over long periods of time. This test is typically done at a 125°C and at some value of V_{CC}

at or above the normal operation specification of the part. It is intended to check for wearout mechanisms related to the junctions, metal, and oxides of the device. It is the ELT and its related test that provide the failures-in-time numbers that were discussed previously. Most manufacturers will tell you those numbers for a base process without any problem. ELT may also catch bond problems as described in Chapter 3.

Temperature cycling

Temperature cycling checks the ability of the part to withstand variations in temperature. Typical temperature cycling tests the part from -40°C to +125°C or from -55°C to +150°C over a cycle of approximately 90 min; these are explained in Mill 883 condition C qualification tests. Temperature cycling exercises the thermal expansion characteristics of the device. The silicon, aluminum, and polysilicon used for interconnection within the die, the passivation along with the plastic encapsulation or the ceramic seal of the package, and the metal lead frame are stressed. Figure 11-4 shows a cross section of the device including passivation layers, metal layers, polysilicon, interconnect, die attach materials, lead frames, and plastic. The expansion coefficients for almost all materials are slightly different, and so mechanical stresses occur.

Notice that on the sides of the die, expansion coefficient differences between silicon, aluminum, oxides, and plastic would result in stresses on the materials that would eventually show up as cracks. These would be failures in temperature cycle. Failures for temperature cycle are found in electrical testing after the temperature cycle.

Temperature and humidity test

Moisture-related testing can be of several kinds. Two atmospheres of fully saturated steam is one condition, high accelerated steam testing is another, 85°C and 85% relative humidity is another, and 85°C and 81% relative humidity is yet another. These tests represent methods of accelerating the moisture susceptibility of the die and package combination, and the impact of moisture in a high-humidity environment. Most plastic packages do allow some penetration of moisture through the package, and that moisture can impact the ability of the device to withstand long periods of operation. The oxides and passivation layers of typical MOS processes use glass with a phosphorus content in it. If there is sufficient moisture in the plastic package the phosphorus forms a phosphoric acid which will corrode both the glass and the aluminum. This corrosion then causes eventual opens, or ionic contamination, and eventual failure of the

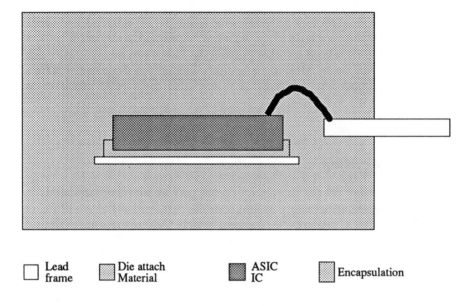

☐ Lead frame   ▦ Die attach Material   ▨ ASIC IC   ▨ Encapsulation

Figure 11-4. Cross section of plastic device.

device. Steam and humidity testing are meant to accelerate those failure mechanisms.

Bake

The final type of stress that is typically done for qualifications of the device is a bake test. Bake consists typically of a high-temperature bake in a ceramic or plastic package. Bake is used to accelerate certain types of failures, including contaminations, such as sodium on the die, which would cause a threshold voltage shift and a major change in performance of the part and probably a failure. Typical devices can bake for a significant amount of time at far above the maximum storage temperature allowed in the specification. This has the goal of then accelerating out all failures and causing any mobile ion contamination to affect the reliability of the part. It should be noted that dynamic burn-in and ELT also accelerate mobile ion contamination. In some cases with an electrical field, contamination can be accelerated during burn-in. A bake may cause ELT contamination failures to recover. This is one of the screens available that would differentiate between a true ELT failure of oxides and a contamination failure caused during burn-in. If the device recovers at the end of a bake, it is probably mobile ion contamination and not a failure of oxides or junctions.

Mechanical stresses

In addition to the electrical and die-related stresses cited in the previous section, there are mechanical stresses that accelerate failures for package-related failure modes. Tests such as solderability, lead strength, mark permanency, thermal shock, constant acceleration all fall into the category of package-related mechanical stresses. These failures are meant to weed out marginally manufactured packages and die that may cause failures in the device.

Many users would like to see qualification data on their own device. This is a very expensive proposition when done on an ASIC device. If there is a need for something in the way of special testing to prove stability of some parameter or to check the sensitivity of one parameter versus another, one could do a subset of a qualification. Doing a full qualification of a device and all the stresses outlined in military standard 883-C is a very expensive process, and it could easily take up to 6 months. Most manufacturers will not recommend doing a qualification of your ASIC.

COST IMPACT

The impact of system of cost must once again be evaluated relative to all the stress testing and qualifications that will be done. A base NRE was quoted for the design, and it probably did not include a qualification. In addition a cost was quoted for the individual device based on die size or gate count the manufacturer expected.

The manufacturer may quote the additional charges for items such as burn-in. Assuming a forecast for the volume of the device that is going to be purchased, it is easy to generate the curves of the various manufacturing qualification options versus total cost. The cost of reworking the design and the cost of failures has to be analyzed on a user-by-user basis.

One estimate that is often cited is a 10-fold increase every time the device moves through the manufacturing process. Therefore finding the failure at the semiconductor vendor is one-tenth as expensive as finding it in a PC board. Finding it on a PC board is one-tenth as expensive as finding it in the system, and so on until it is one-tenth as expensive as finding it in the field. Noting the failure rate as quoted by the manufacturer for infant mortality and long-term failure rate, one may derive different curves for the cost of manufacturing and total system cost for the device. This would be helpful in trying to analyze where one should draw the line relative to additional expenses. The quality level of

Table 11-1. Cost with and without burn-in.

PROCESS STEP	ASIC WITHOUT BURN IN	ASIC WITH BURN IN
ASIC Cost	$5	$7
PC board cost	$200	$202
Debug cost	$55	$50
PC board cost	$255	$257
System failure	0.1%	0.05%
Replacement cost per board	$25	$12.85
Total cost	$280	$269.85

semiconductors as manufactured over the past several years has improved dramatically. The incremental improvement in performance that is received by additional stress such as burn-in and additional testing that are added to the manufacturing process may not be worthwhile. Designers must make their own calculations as to whether it is worthwhile to spend money to improve the reliability and quality of their devices.

Once the assumptions to do the calculations are made, the numbers will speak for themselves. Table 11-1 shows a typical calculation for total cost of the device with multiple parameters. Numbers with and without burn-in are shown. Those numbers are compared to cost of reworking the system and the approximate failure rates.

The next set of curves is the same relative criteria of gate array or standard cells with or without burn-in versus volume production (see Figure 11-5). The relationship of the two curves in a particular design would be one of the criteria for choosing any additional work that may or may not be implemented on the device.

One type of calculation that can be done is a total systems cost calculation. This calculation entails the cost of the components and the cost of the extra processing versus the probability of a failure. Table 11-1 is a typical calculation of total system cost for two different devices with slightly different failure rates and different PC board and repair costs. Notice that, depending on the component cost, there are times when it is worthwhile to burn in and other times when it is not.

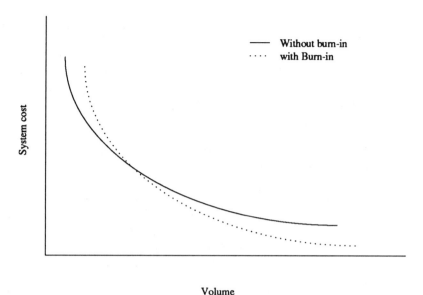

Figure 11-5. Volume impact on burn-in cost.

THE IMPACT OF TESTING ON QUALITY AND RELIABILITY

The book so far has covered the reliability and quality impacts of devices and the cost of typical types of failure. One item that is worth noting at this time is that the design and its margin to specification does have some impact on the reliability and quality of the device. Semiconductor failure modes fall into two categories. The first is a catastrophic failure, such as a defect causing physical shorts and opens in the device. In this case the logic does not operate correctly.

Such a failure can almost always be screened out with a reasonable test coverage in any of the functional patterns and in general do not change over time. In general, most defects of this type are in the multimicrometer size. They include such defects as the bridging of two lines in either silicon or diffusions or metal. This is a sufficiently large failure and is consistently bad.

The second type of failure is a defect such as gate oxide degradation, junction leakage, crystal lattice imperfections, threshold shifts, and resistive contacts. These defects have a tendency to degrade or change characteristics over time. In some cases they stabilize, and in other cases they continue to get worse. Hot-electron problems and charge trapping in oxides have a tendency to continue to degrade over the life of the device. This degredation causes a shift in the parameters of the part over extended operation.

Figure 11-6 shows the effect of a simple two-invertor circuit with and without a leakage path to ground and the appropriate change in the speed of the part. Notice, as the leakage currents increase, that the speed of the part degrades slightly. If the leakage is induced by gate oxide and accelerates or changes over time, it would produce a slower and slower device relative to the specification of the invertor. Oxide charge trapping has a tendency to do exactly the same thing, as shown in Figure 11-7. This figure shows the threshold voltage of the device versus performance, and a very minor change in threshold voltage can degrade the speed of the device fairly dramatically. Most vendors will evaluate items such as V_T stability and leakage currents or junction stability in the qualifications, and that data is available in most cases for review by the customer.

It may not be obvious at this point, but if test vectors are generated with guardbands, as discussed in Chapter 6, minor changes in performance of one NAND gate would not cause failures. For instance, in a cell chain if such a NAND gate is not in the critical path and if there is margin to specification, these shifts will not cause the system to degrade. This type of failure mode is generally represented in one gate or two gates within the circuitry and very easily would not impact the overall performance of the device if there is any margin to spec. Take a look again at a typical system design with 10 combinatorial elements between the latches as shown in Figure 11-8. In this case one gate's propagation delay increases by 2 ns. In the first case where the system spec is tight, it may cause a failure.

In the second case that same degradation would not cause a failure. Therefore, the margin in simulation would improve not only the repeated performance, but the long-term quality and reliability of the device. This is very important for ASIC devices in that it is very difficult to find these type of failure modes. The farther the device is away from the specification need of the system, the better off for reliability. "Farther," in this particular case, is defined as the difference of a few gates. For example, if the design is running at 20 MHz meaning a 50-ns cycle time, it would be allowed twenty-five 2-ns gates between latches (less the latch delay). If there is margin, such as only 23 gates, then there is margin to spec.

If the design could be modified by increasing the parallel logic in the computation to get to the point where there is margin, again reliability is increased. This allows the maximum propagation delay to degrade versus the minimum cycle time of the ASIC. This slight degradation would not have a detrimental impact on the performance of the part. Failures of the kind described in this section, including gate oxide, threshold shifts, and leakage current degradations, usually impact only one gate at a time. So if this is a logic design of 20,000 gates, most critical paths should not be tight relative to specification. The probability of a defect of this kind hitting the gates in the one or two critical paths that have minimum guardband is very small. If the critical paths have margin in them, the probability becomes even smaller. This impact of improving the test

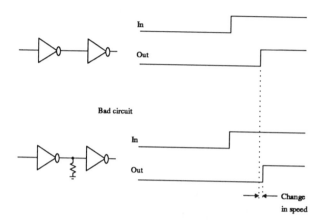

Figure 11-6. Leakage in a circuit.

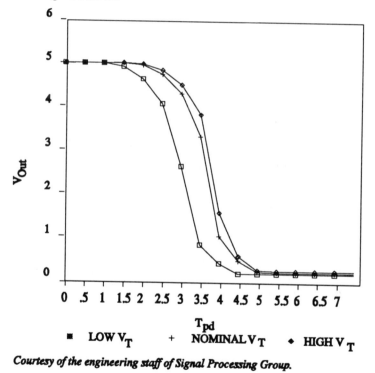

Courtesy of the engineering staff of Signal Processing Group.

Figure 11-7. V_t shift speed impact.

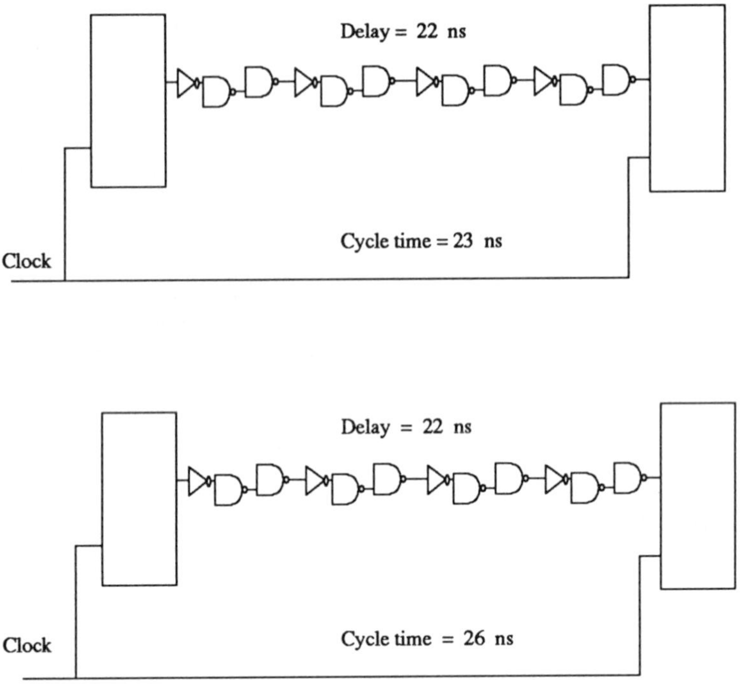

Figure 11-8. Guardband impact on the critical path.

coverage and overall speed of the part relative to specifications will give more margin for error and a more stable higher-quality system.

SUMMARY

The impact of testing on quality and reliability can be dramatic. System failures, and quality problems can be minimized if a comprehensive high-quality testing program is developed for the ASIC device. If little attention is paid to testing, quality and reliability problems may become significantly worse.

12

SPECIAL CONSIDERATIONS

One of the fastest growing segments of the ASIC industry is the mixed-mode, or mixed-digital, analog ASIC design. Libraries of many manufacturers now support basic functions such as small A-D converters, D-A converters, comparators, and other linear functions. Unfortunately, optimizing for digital logic speed leaves little operating range for analog. Some libraries by specialty manufacturers are primarily oriented around analog.

ANALOG LIBRARIES

Analog libraries may have op amps, resistors, capacitors, comparators, and all the other linear building blocks necessary to build an analog system. If the end system has analog components in it, it may be worthwhile investigating the addition of analog to the ASIC device. Figure 12-1 shows a typical analog system with a combination of digital logic and discrete analog functions integrated into a single integrated circuit. Depending on the type of analog that is installed on the ASIC integrated circuit, many additional considerations may be taken into account, the foremost of which are grounding and power considerations for the two portions of the integrated circuit. Analog is very sensitive to the noise of digital systems, and most manufacturers will partition the device and actually require the grounds to be separated. Power should be separated if the analog is of reasonable performance. Building an analog system in the schematic capture mode is very similar to that of the digital system except that the building

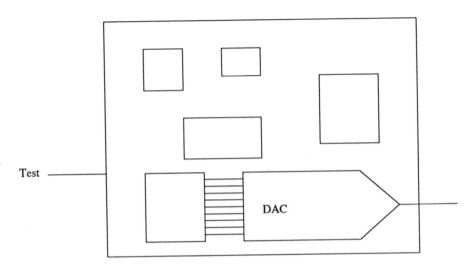

Figure 12-1. ASIC with analog cells.

blocks might be larger. Working with predefined analog blocks such as A-D, D-A, and comparator type of circuitry is easy.

Some manufacturers will allow discrete transistor-level analog design to be implemented. This will allow the designer to work with p-channel and n-channel MOS transistors. The simulators for analog circuitry over the past several years have made significant progress. It is now quite easy using spice simulators to simulate a large analog system for performance in the manner similar to what would happen in a digital system. Some of the more advanced simulators also allow the linking of digital and analog function. They can also handle the modeling of external components such as motors and input sensor devices. These linear circuit design techniques allow for designers to create complex logic functions and complex analog functions on one integrated circuit. The designer can do it without having to worry about the interaction and performance of the systems separately. It is estimated that this portion of the market will more than quadruple in the next several years. As much as 40 percent of the ASIC market will use analog by the mid-1990s.

NOISE IN ANALOG AND DIGITAL SYSTEMS

One of the main considerations one should look at in doing a mixed analog and digital system—beyond the issue of simple noise and decoupling within the system—is the unique aspect of testing. Testers that accurately test analog functions are usually different from the machines that are commonly used to test the digital functions. There are mixed-mode testers commercially available that can be purchased for analysis, debugging, and production manufacturing of mixed-signal type devices. Depending on the complexity of the analog circuit that is being designed, the circuit may not require a vendor to have mixed-signal test capability. For instance, if designing items such as filters and companding DACs, you will probably need a special piece of test equipment to test the part. This equipment would accurately measure, to a significant number of bits the performance of the system.

SELF-TEST FOR ANALOG

Building in self-test capability for linear integrated circuits is also a very important aspect. Adding self-testing and digital sequence controls for analog circuits becomes a fairly complex task for large designs. Perhaps you should reduce some of the logic around it to a behavioral model so that the functionality of the analog can be accurately analyzed. Doing spice simulations on the analog function allows accurate determination of step function response and pole and zero analysis. These simulators and testers may also have the capability to do frequency and time domain analysis, but they are unavailable in present digital circuit simulation programs or digital testers. Figure 12-2 shows a simple analog system designed with a combination of digital and analog functions.

Figure 12-3 shows the appropriate test sequence to analyze the functionality of the part in a design verification system. For testing A-D and D-A converters,

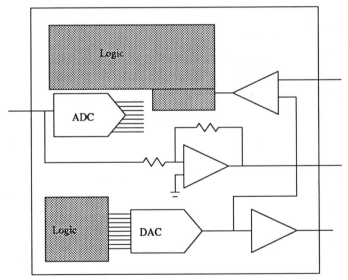

Figure 12-2. ASIC with complex analog.

measurements of the steps, the linearity, and the monotonicity of the device is one of the important aspects of testing analog circuits. Figure 12-4 shows the linearity test for an analog integrated circuit, with the input value on one axis and the output value on the other axis. An ideal linear circuit would be straight-line, but because of the fact that it is a digital system, there are step functions. Depending on the accuracy and bit placement, the measurement system must differentiate certain types of linear errors. These typically require mathematics to be performed.

When doing this in a test system not designed to do linear testing, the starting and stopping of the tester to do the math can consume considerable amounts of time. The large ATE systems used by semiconductor companies that have this capability allow these measurements to be done on a very rapid basis.

EPROMS

Some manufacturers now supply EPROM process technology in their ASIC libraries. The design can be done with not only the common AND/OR gate logic used in most digital systems, but a small or medium size EPROM can be added to the device and programmed. This EPROM can be a straight ROM type of function or can actually be a programmable logic device implemented with EPROM technology.

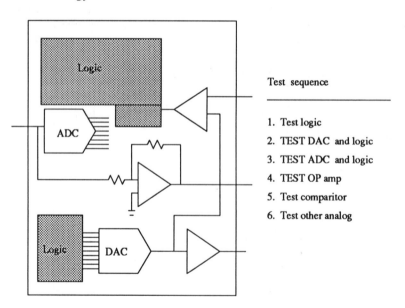

Test sequence

1. Test logic
2. TEST DAC and logic
3. TEST ADC and logic
4. TEST OP amp
5. Test comparitor
6. Test other analog

Figure 12-3. Analog test sequence.

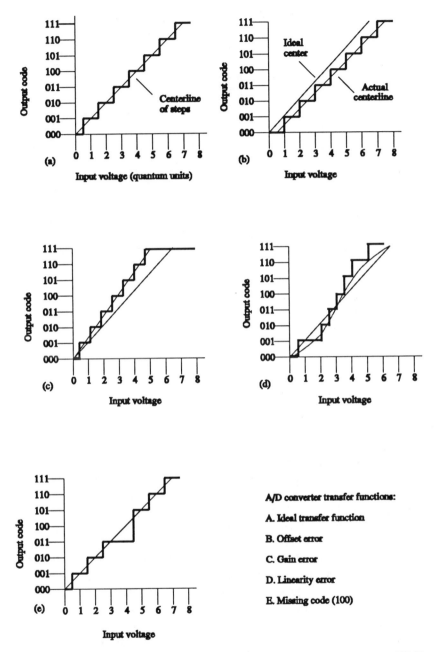

Figure 12-4. A/D linearity test. *Courtesy of R. Feugate, and S.McIntyre*, Introduction to VLSI Testing, *1988, p.161. Reprinted by permission of Prentice Hall, Inc., Englewood Cliffs, New Jersey.*

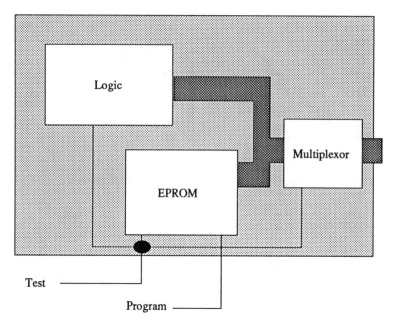

Figure 12-5. EPROM in an ASIC.

The testing of EPROM cells in a manufacturing flow is fairly difficult. If the blocks that are being used are standardized and provided by the manufacturer, implementation of them in the system is not a very difficult task. For instance, Figure 12-5 shows a logic design with an embedded EPROM. The manufacturer's test program would look much like that in Figure 12-6; the entire logic functionality of the part should be checked. Then a special flow for reading of the blank EPROM, programming a data pattern into it, and checking that the data pattern is still valid is then added to the test program.

These EPROMs should have test logic added to them so that the functionality of the system could be checked without having to program the individual EPROM device—much the same way as a ROM would be checked for its data contents and not its functionality within the system. The next step in the EPROM manufacturing process is called a *retention check*. This is where the pattern that is programmed into the EPROM is baked at a high temperature, or stressed, to ensure that the charge is stable on the gates of the device. This ensures that the gates or cells of the device do not lose data over an extended period of time or when exposed to high temperature.

Another test is then added to the manufacturing flow that verifies that the data pattern programmed is still available in the part several days later after high-

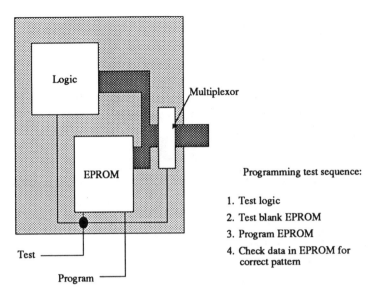

Programming test sequence:

1. Test logic
2. Test blank EPROM
3. Program EPROM
4. Check data in EPROM for correct pattern

Figure 12-6. EPROM test sequence used in the test program flow.

temperature storage or stress, as shown in Figure 12-7. This ensures that the device is stable and that leakage paths are not affecting the data retention of the device.

It should be noted that the total charge stored on the EPROM gate is somewhere in the range of 100 femtocoulombs. If the manufacturer has predefined EPROM logic blocks that can be added to the circuitry and the manufacturer treats testing in a similar manner to logic blocks such as microprocessors and other large functions, the test is not a concern. If the manufacturer takes full responsibility for testing of these logic blocks or EPROM blocks, there is very little to worry about relative to implementing EPROMs as a portion of the design. If the manufacturer does not take the responsibility for testing, this is a new set of problems, and it may be worthwhile to avoid. Implementing of the flows previously described for EPROM verification is a fairly difficult process to do and can be fairly expensive. In addition, if the manufacturer allows EPROMs to be placed within the circuitry without predefined functions or connection points for programming, the problem of defining test sequences becomes even more pronounced. Adding this capability to the integrated circuit is a very difficult thing to do and should be left to individuals who are experts in the field.

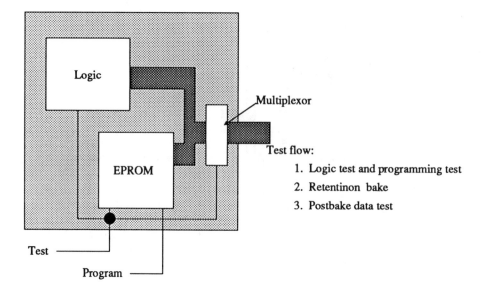

Figure 12-7. EPROM test flow (steps in the manufacturing process).

EEPROMS

In addition to the EPROMs that are available in some libraries, EEPROMs, or E^2PROMs, are now available in libraries supported by some vendors. E^2PROMs are actually somewhat easier to test because programming the device is generally done at lower voltages and because erasure can be affected in the system very rapidly. This is contrast to EPROMs which need ultraviolet light for erasure as in an EPROM technology. However, this ease of testing and erasing becomes a problem for programming. This is because of the need for selective pulse widths and flows to verify not only functionality, but data retention of the device. Once again, however, if the manufacturer takes responsibility for the test generation of these devices, then proceed. If the manufacturer does not, it would be worthwhile to analyze completely the responsibility of the manufacturer for generation of the E^2PROM device test patterns.

Figure 12-8 shows a typical E^2PROM system design that includes a small section of EEPROM, or E^2PROM. This E^2PROM could be used to update tables such as prices in a cash register or seasonal parameters that don't change too frequently. Another example is the set up of pc systems memory configuration. Designing and implementing a circuit like this is a relatively easy task. The manufacturing of it may pose significant problems unless treated with sufficient detail for the both the testing and manufacturing related issues.

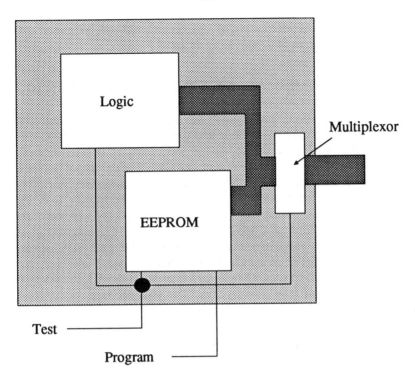

Figure 12-8. EEPROM in an ASIC.

ECL

Most of this book has focused on CMOS technology. There are other processes available. ECL, for instance, is the highest-volume, highest-performance process presently available, and there is a tremendous amount of gate array activity supported in ECL designs. Most bipolar processes, ECL in particular, lend themselves to gate arrays.

There are no known standard cell libraries available for the ECL processes today. In consideration of ECL it is important to note that the power of the device is significantly greater than that of MOS devices, but it is much less speed-sensitive. MOS devices generally involve ceramic packages with heat sinks for fairly large devices. The power dissipation of ECL does not vary with the speed of the integrated circuit as much as in MOS. The number of manufacturers of ECL-type gate arrays is significantly less than that of the MOS manufacturers for standard cells and gate arrays. Device design considerations and testing considerations are of equal importance. The ability to do an ECL design is as easy as a MOS design, and one should not be intimidated by attempting to generate an ECL gate array design.

GALLIUM ARSENIDE (GAAS) ASIC SEMICONDUCTOR

Gallium arsenide capability has come of age in the last few years. Cell libraries are now available in gallium arsenide and commercial systems are made with gallium arsenide. Gallium arsenide has the performance characteristics of very high speed and yet has some of the features of CMOS in its ease of implementation. Gallium arsenide technology allows the implementation of 1K RAMs, for instance, that have propagation delays of just a nanosecond or two. This is significantly faster than what can be done in a CMOS type of logic design. Designs that are planned in a gallium arsenide can be implemented in an integrated circuit today. This can be either a standard cell or gate array format. It is essentially no different from standard CMOS design. Note, however, that the speeds are significantly faster. Gate delays may be in the 50- to 100-ps range and power dissipation may be very low. The tradeoff here is that cost per function is significantly more than in a typical CMOS gate array or standard cell library. Figures of 10 times the CMOS cost are not uncommon.

Several manufacturers make gallium arsenide devices, and the list is significantly smaller than that of the CMOS or ECL standard cell and gate array vendors, but they do exist. Although the price is significantly higher, the design process and testing considerations are exactly the same. These libraries are usually about an order of magnitude faster in performance than CMOS. Testing of the devices, therefore, becomes an even greater problem. The commercial testers that were referenced in Chapter 2 have typical accuracy specs in the 0.5- to 1.0-ns range. For gallium arsenide integrated circuits with propagation delays of 100 ps, the tester accuracy must be improved dramatically, or ac testing must be done in a different manner.

To test these devices, high-speed techniques such as frequency counter and sampling oscilloscopes and high-speed pulse generators must be used for propagation delay testing. It is not uncommon to have system specifications in the 100- to 200-ps level. This is beyond the capability of all but the most expensive commercial ATE systems. When trying to make these kinds of measurements on a typical commercial system, the accuracy of the machine must be in the 100-ps range or less to be accurate. A large pin-count tester for testing gallium arsenide or high-speed devices in this speed range, is generally a multi-million-dollar-plus commercial test system.

In addition, during testing for these devices at 100 ps of propagation delay, lead lengths of 1-in. start degrading the performance and the measurement accuracy. And so one needs to pay attention to the physical connections of the integrated circuit during its final testing period.

BICMOS

BICMOS processes have the attribute of combining the low-power and high-density characteristics of CMOS with the high-speed and high-drive capability of bipolar devices. The process actually fabricates both kinds of devices in the masking sequence. In a typical application, npn transistors can be fabricated for outputs that have extremely high drive capability relative to their CMOS counterparts. Figure 12-9 shows a cross section of a BICMOS device. The implementation of gates within the device consists of a typical CMOS invertor, NAND, AND, NOR, OR structure followed by an npn totem pole. This is similar to the output arrangements of a TTL integrated circuit. See Figure 12-10 for a typical NAND gate structure.

Optional methods of implementing BICMOS include implementation of logic internal to the arrays and internal logic that is strictly CMOS and the use of the npn transistors in the BICMOS structure only for outputs. Using the stacked approach, the CMOS and npn gate model is the fastest speed but of a lower density than normal CMOS. Using the CMOS array with bipolar output capability allows implementation of dense CMOS arrays with bipolar drive capability at the outputs.

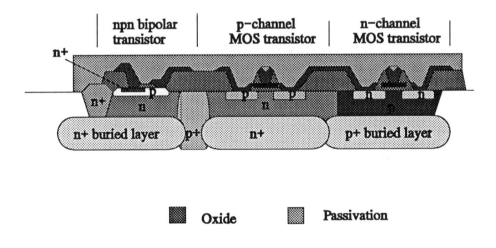

Courtesy of the IEEE International Solid State Circuits Conference © 1986.

Figure 12-9. BICMOS cross section.

BICMOS processes are more complex than their counterpart CMOS processes due to the additional masking steps in the fabrication process, the processes are more costly, but the devices are faster.

RADIATION HARDENING

For military and space applications, radiation hardening services are now available through the semiconductor vendors that supply the devices and through outside services that perform the same radiation hardening service. Radiation hardening is a very small field, but it is commercially available. Contact the semiconductor manufacturer or foundry to find out what is recommended for radiation hardening of the integrated circuits.

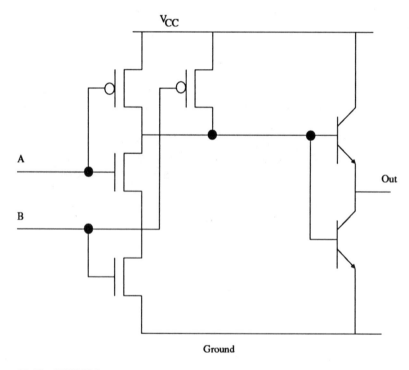

Figure 12-10. BICMOS gate structure.

CONCLUSION

Many aspects of ASIC device manufacturing are impacted by testing. The inclusion of testability circuits in the design of the ASIC integrated circuit helps make the part understandable in the debugging process. What's more, good testing and design practices will make the device easy to manufacture.

GLOSSARY

Portions of the following reprinted with permission of IEEE, Design and Test of Computers, *1989, IEEE.*

Adapter—The part of the load board of a tester which allows the wiring of pin electronics channels to connections on the device under test.

AQL (acceptable quality level)—An outgoing inspection sampling plan used by some manufacturers to verify quality levels.

ASIC (application-specific integrated circuit)—This is a device designed for a sole application, usually by the user.

Behavioral model—A model of a block of logic which acts correctly as seen by the outside of the block, but does not contain all the internal logic gates. This speeds up simulation time considerably.

BICMOS—A process consisting of CMOS devices and bipolar transistors. This process gives the speed of a bipolar device and the power of CMOS with a minor density tradeoff for the additional npn transistors. This process also includes several additional masking steps and may cost substantially more than a CMOS process, but speed is closer to that of ECL.

BIST (built in self test)—This allows the device to test itself by supplying input patterns and comparing output results of patterns.

Boundary (or external) scan path—A scan path formed by associating a bistable with each of the I/O bonding pads; used in testing to control or observe signals on the IC pins.

Built-in test—Sometimes used as a synonym for built-in self-test. Also used to specify the execution of a diagnostic program or the inclusion in the product of circuitry to detect the occupance of a run-time error (concurrent on-line checking).

Built-in self-test (BIST)—Capability of a product (chip, multichip assembly, or system) to carry out a functional test of itself. Some support from external equipment may be required. BIST usually involves special hardware in the product to generate test inputs and to analyze test responses.

Characterization—The testing of the device or process at or outside all performance corners, speeds, and voltages. Also defined in some cases as simulation for performance of the devices at its operating extremes.

Comparator—The circuit in the tester that checks expected data and voltage versus actual data.

Continuity—The checking of connections from the tester to the DUT.

Controllability—The ability to control a circuit for test purposes.

Correlation—The testing of a sample of devices to ensure that the functionality in the test system is identical to operation in the application. This step would also include a detailed analysis of failures. It is also used to check one set of test patterns against another.

Defect—Any manufacturing aberration that causes the device to be nonfunctional or nonreliable. Defects can be caused by particles, scratches and handling, or impurities in the materials used in the process.

Die—The basic ASIC device in its smallest from. May also be called a chip, device, slice, or unit.

Diffusions—The portion of the semiconductor processing that changes the dopant concentration from p type to n type or vice versa.

DPM (defects per million)—A typical measure of outgoing quality.

DUT (device under test)—The device being tested in the test system.

Dynamic—Logic or memory circuitry that is designed to store charge on capacitive nodes. This technique allows small densities but also has the problem of nodes discharging which therefore need to be refreshed on a regular basis. The major problem with testing dynamic circuitry is that slow-speed analysis and looping may not be possible. Slow-speed high-temperature testing are usually needed to find leakage failures.

EEPROM—This is an electrically erasable PROM, usually formed by a floating gate in the device.

ELT (extended or elevated life test)—A long-term burn-in of the part to find the failure rate.

Encapsulation—The process of sealing the device during assembly, in its final package.

EPROM—This device is made using a floating gate which must be erased with ultraviolet light. The need for ultraviolet light is what differentiates it from the EEPROMs, or E^2PROMs. Programming for the two is roughly the same, and testing entails approximately the same activities.

ESD (electrostatic discharge)—This is the buildup of charge on a body when moving across an insulated surface. When coming in contact with an integrated circuit, ESD may cause extremely high voltages and currents for very short periods of time. These high voltages and currents may damage the input protection of the device.

Exhaustive testing—functional testing in which all possible input combinations are applied to a combinational circuit, or a checking experiment is applied to a sequential circuit.

Fabrication—The process of manufacturing the integrated circuit. Depending on the vendor used, the company may reference fabrication as wafer fabrication, package fabrication, or sample fabrication.

Fault coverage—Figure of merit for a test procedure: the fraction of all possible faults in a particular class (almost always single stuck-at faults) that the test procedure has detected.

Fault grading—A metric of the quality of the test program, usually expressed as a percent (such as 97%, 99.4%, 99.99%). This is calculated by dividing the total caught faults from the test patterns by the number of possible faults times 100.

Faults—Errors, or stuck-at conditions, impressed upon the device during simulation for testing to ensure that errors in the device or defects made during fabrication will be caught using the test patterns of the device.

FIT (failures in time)—A failure rate used to describe failures in the life of the device. An inherent failure rate number.

Formatting—The modification of the data patterns used for simulation and the clocks in the tester to look at selected timing conditions.

Foundry—The vendor, or supplier, who manufactures the wafers and package devices for the designed ASIC.

Functional (boolean) testing of a digital product—A sequence of valid input signals is applied to the device under test, and the output response is compared to the correct response. Only the logic values represented by the signals (not the exact voltages or currents) are considered. Valid input signals are signals within device specifications. Discussion: In connection with chip tests, the functional board test, a functional test applies test signals to an entire board (or large portions of a board) using board-edge connectors (and perhaps some special test connectors). The other class of board test—in-circuit test—tests individual ICs on the board using "bed of nails" fixtures. There is another use (or misuse) of the expression "functional testing." It is sometimes used to describe a test that detects failures by "verifying correct operation" rather than by verifying the absence of specific faults. This usage is based on the desire to generate test patterns without reference to specific implementation or to a model of which faults are detected by the test.

Functionality testing—Verification of the device for its logic correctness. This is different from parametric or performance testing of the device where ac or dc parameters are checked. In a functionality test only the truth table, the ones and zero input and output conditions, are checked for the integrated circuit.

GAAS (gallium arsenide)—This is a process using the compound gallium arsenide instead of silicon as the base material for the semiconductor process. Dopants, oxides, and metals are used. Gallium arsenide has the characteristic of being faster than silicon. Processing is less advanced. Costs are higher.

Galpat—This is a galloping pattern which allows the functional testing of RAMs for all possible address transitions. It is one of many possible patterns for testing RAM data for row and column address sensitivity.

Guardbanding—The process of adding margin between the device simulation and test values to ensure repeatability and, to some extent, reliability.

Hierarchical design—Design techniques that start at a top level of logic and work down to detail block implementation.

I_{OH}—Output drive current of a device in the high state.

I_{OL}—Output drive current of a device in the low state.

JTAG (Joint Test Action Group)—A collaborative organization composed of major semiconductor users in Europe and North America. Author of a proposed standard for a boundary scan register and test access port (TAP), which has evolved into the proposed IEEE standard P1149.1.

$L_{effective}$—The channel length of an MOS process.

LFSR (linear feedback shift register)—A shift register with connections from some of the stages to the input of the first element though an exclusive-OR gate (standard form) or through a shift register with connection from the last stage to exclusive-OR gates at the inputs of the first and intermediate stages of the register (modular form).

Lithography—The photolithography process used to pattern wafers for etching, doping, or oxidation. The lithograph process determines the minimum possible geometry and therefore the speed, density, and performance of the process.

LSSD (level sensitive scan design)—This is also called *scan design*. It is a technique where all storage elements in a device are chained together in a dual mode. The first mode being the normal operation of the device where clocks allow the storage of data in normal system operation. In the second mode of the device, clocks are used to shift data in and out of the device for testing purposes.

Macro—A series of cells connected together to make a larger block of logic. Macros can be any level of complexity from simple latches, or macro consisting of two cross-coupled NAND gates, to a microprocessor. They may be hard or soft. A hard macro is a fixed layout; a soft macro varies with placement.

Microprobing—The process of using very fine probes to force inputs into, and measure the logic state output of the device; also the process of analyzing the device by physically contacting internal nodes in the device. The equipment required to do this requires the placement of probes on metal lines, usually only a few micrometers in width. Microprobing is one technique for contacting the internal nodes of the device to ensure that the logic is functioning correctly. Other techniques for measuring internal voltages or states include voltage contrast electron microscopy, and test access schemes.

Monotonicity—The characteristic of an analog to digital converter that shows consistently increasing digital readings with a continuously increasing analog reading. Correct monotonicity of the device would ensure that each increasing analog value would relate to a correspondingly higher digital value read at the outputs.

Net list—The listing of the interconnections of the device. This includes not only all the AND/OR gate functions used, but also the interconnections of blocks. This may include fanout and race conditions also.

Nonrecurring engineering charges (NRE)—The charges for one time setup for manufacture of the ASIC. This is usually associated with the mask, probe card, wafer manufacturing, package assembly, and testing of the prototypes.

Observability—The ability to observe nodes internal to the part.

Oxide—Primarily referred to here as silicon dioxide (SiO_2), which is a primary insulator in the semiconductor process.

Parallel signature analyzer—An LFSR used to compact several test-response signals from the circuit being tested. The test-response signals are connected to the inputs of individual LFSR stages through exclusive-OR gates: also called an MISR, or multiple-input signature register.

Parametric test—electrical test that evaluates parameters such as dc electrical characteristic (V_{IH}, V_{IL}, I_{IN}, V_{OH}, etc.) and ac electrical specifications (timing parameters).

Parasitic—Any element formed in the manufacturing process which is not directly related to the logic. Parasitic can be resistance and capacitance in the interconnection and can also be the formation of leakage paths due to contamination and processing errors.

Passivation—Final covering of the IC to protect the circuit below from scratches, contamination, and moisture.

Pellicle—A clear coating over the photoplate that prevents dust from landing on the focal plane; thereby prevents printing the image of the dust on the wafer.

PLA (programmable logic array)—An array of devices that performs combinatorial logic.

PLCC (Plastic Leadless Chip Carrier)—A Jedec standard package which includes wrap around leads, commonly referred to as J bend.

PMU (parametric measurement unit)—This is the portion of the integrated circuit tester that does the accurate measurement of dc currents and voltages. PMU ranges typically are from less than a few nanoamps to as much as 1 A in dc current measurement. Voltages from 1 mV to approximately 10 V are typical. Most PMUs are accurate to three digits or more of accuracy. Connection of the PMU to the device under test is usually done via a matrix, either relay or solid state. This prevents the PMU with its high capacitance and associated loading from affecting the device during functional testing.

PQFP (plastic quad flap pack)—This is a 25 mil pitch jedec standard or a 0.6-, 0.8-, or 1.0-mm EIAJ standard package. This particular package has a gull wing lead configuration that allows direct surface mounting of the package on the printed circuit board.

Prototypes—The first shipments of an ASIC device. These may not be fully manufactured or tested to specification. Prototypes are typically rush-manufactured and are used for evaluation of the functionality of the silicon.

Pseudo-exhaustive testing—Functional testing in which exhaustive test inputs are applied to segments of the circuit being tested.

Pseudo-random testing—Testing in which pseudorandom binary numbers are used as test inputs.

Pseudo-random binary numbers—Binary numbers that are generated by a logic circuit or algorithm. The numbers have certain randomness properties but are reproducible. No numbers appears a second time until all possible numbers have been generated. Discussion: Pseudo-random binary numbers are most often generated using an LFSR. Such a circuit does not generate the all-zero binary number. All possible numbers for an n-stage LFSR are n-binary numbers except the all-zero number. No number repeats until all non-zero numbers occur.

QA (quality assurance)—The step in the manufacturing process that verifies performance to specification. Also checks that all testing and manufacturing steps were done correctly.

RAMs (random access memories)—This is a data storage element. RAMs allow the easy changing of data patterns and can typically be written and read in a few nanoseconds.

Random testing—Testing in which random binary numbers are used as test inputs.

Reliability—Long-term stability and continuous functional operation of the part.

ROMs (read only memory)—ROMs are typically mask-programmed at the factory. Data contents are supplied either in the form of a net list or in a data pattern supplied to the vendor for the programming of the ROM.

Scan path—A shift register made up of the system bistables. In test mode, the bistables are connected in a shift register, during normal operation they carry out their normal system functions. The scan path is used to shift test data into the system bistables and to shift out test response data.

Self-test—Same as built-in self-test.

Serial signature analyzer—An LFSR used to compact a single test-response signal from the circuit being tested. The test-response signal is connected to the input of the LFSR through an exclusive-OR gate. The final contents of the LFSR is called the signature.

Simulation—The process of using an engineering workstation or computer to simulate the functionality of the device. Simulation can be the digital simulation (functionality) or performance simulation (timing, drive, levels).

Single stuck-at fault—A fault in which one line in the circuit has a fixed signal value (representing zero or one) independent of other signal values. Discussion: Some fault simulators permit stuck values only on gate outputs and not on gate inputs. This is the same as assuming that the fanout branches cannot have stuck values independent of one another and that we need to consider only fanout stems. Clearly, there is a loss of accuracy in using this gate-output stuck-fault model.

Specifications—The parameters and functionality guaranteed by the vendor for the device purchased. This includes ac, dc, and functionality parameters.

Synthesis—This involves using a high-level language or equations to generate logic versus entry of schematics.

T₀—The start of the tester period within all test systems. This is the time that all timing generators and edges are referenced to. All comparisons take place relative to T_0.

Test-response compaction—Reduction of the test-response data stored or observed; usually done with a signature analyzer. Discussion: Recently, some authors have started using compaction. They are not the same. In compaction, information is lost, while in compression, redundancy is removed without the loss of information. Using the two synonymously is not recommended.

Testability—The ability to control and observe the part in the tester.

Tester—The machine that allows the accurate testing and debugging of the device.

Testing—The process of determining whether a piece of equipment is functioning correctly or whether it is defective. Equipment can be defective because it doesn't function as designed or specified.

TPT (throughput time)—This is the total amount of time from submissions of the net list or design to the vendor, until samples are delivered. Throughput time may also be defined as the time from receipt of the order to production shipments.

Verification—Proving that the design works correctly. This is done by methodically going through the device and verifying that the logic functionality and performance of the integrated circuit is as expected by the simulator and specifications.

VHDL (VLSI hardware description language)—This language is a standard proposed by the IEEE; it allows for the definition of logic in a high-level language. See also synthesis.

V_IH (voltage for input level highs)—This is a parametric value to be used in testing. It must relate to the output levels of the device driving it less a guard band.

V_IL (voltage for input level lows)—Also used in the specification for the IC.

VLSI (very large scale integrated circuits)—VLSI devices usually contain 100K or more gates.

V_OH—Output high voltage of the part. Typically tested during the dc performance testing of the part.

V_OL—Output low voltage of the part.

Voltmeter—Any instrument used to measure the voltage on a particular node. Test systems may include voltmeters or PMUs for the accurate dc measurement of parameters.

Wafer—A round substrate used for the manufacturing of the integrated circuit typically made of silicon for most MOS and bipolar processes.

SUGGESTED READINGS

Acken, John Michael. *Deriving Accurate Fault Models.* Stanford University; Computer Systems Laboratory, 1989.

Agrawal, Vishwani D., and Sharad C. Seth: Fault coverage requirments in production testing of LSI circuits, *IEEE Journal of Solid-State Circuits,* vol. SC-17, no. 1, February 1982, pp. 57-61.

—Agrawal—: *IEEE Computer Society. Tutorial; Test Generation for VLSI Chips.* Washington, D.C. : Computer Society Press, 1988.

Bennetts, R. G.: *Design of Testable Logic Circuits.* Reading, Mass.: Addison-Wesley Publishing Company, 1984.

Carr, William, and Jack Mize: MOS/LSI Design and Application, New York: McGraw-Hill Book Company, 1972.

Cole, B.: Making the right moves in ASICs, *Electronics,* November 1989, p. 56.

Comer, David J.: *Introduction to Semiconductor Circuit Design.* Reading, Mass.: Addison-Wesley Publishing Company, 1968.

Conway, L. and C. Mead: *Introduction to VLSI Systems.* Reading, Mass.: Addison-Wesley Publishing Company, 1980.

Crosby, Philip: *Quality Is Free.* New York: McGraw-Hill Book Company, 1979.

De Bono, Edward: *De Bono's Thinking Course.* United Kingdom: Facts on File Inc., BBC, 1982.

Electronics Engineering Times. Standards, software lend design status to test, *Electronics Engineering Times,* August 21, 1989, p. 80.

—. Credo for a new decade: design for testability, *Electronics Engineering Times,* August 7, 1989. p23.

Foster, Caxton C.: *Computer Architecture.* New York: Van Nostrand Reinhold Company, 1970.

Fuegate, R., and S. McIntyre: *Introduction to VLSI Testing.* Englewood Cliffs, N.J.: Prentice-Hall, 1988.

Gise, Peter E. and Richard Blanchard: *Modern Semiconductor Fabrication Technology.* Englewood Cliffs, N.J. : Prentice-Hall, 1986.

Gray, Paul R., and Robert G. Meyer: *Analysis and Design of Analog Integrated Circuits.* New York: John Wiley & Sons, Inc., 1977.

Grove A.S.: *Physics and Technology of Semiconductor Devices.* New York: John Wiley & Sons, Inc., 1967.

Hamilton, Douglas, and William Howard: *Basic Integrated Circuit Engineering,* New York: McGraw-Hill Book Company, 1975.

Harwood, Wallace, and Mark McDermott: Testability Features of the MC68332 Modular Microcontroller. *IEEE International Test Conference Proceedings.* IEEE New York, 1989. pp.615-623.

Hill, Fredrick J. and Gerald R. Peterson: *Digital Logic and Microprocessors.* New York: John Wiley & Sons, Inc., 1984.

—, and—: *Digital Systems Hardware Organization and Design.* New York: John Wiley & Sons, Inc., 1973.

—, and—: *Introduction to Switching Theory and Logical Design.* New York: John Wiley & Sons Inc., 1981.

Hodges, David A.: *Semiconductor Memories.* New York: IEEE Press, 1972.
—: *1937-comp Semiconductor Memories.* New York: IEEE Press, 1972.
ICE Mid-Term Status and Forecast of the IC Industry. Scottsdale, Ariz.: Integrated Circuits Engineering Corporation, 1989.
ICE Status 1988—A Report on the Integrated Circuit Industry. Scottsdale, Ariz.: Integrated Circuit Engineering Corporation, 1988.
IEEE Design and Test of Computers, bimonthly.
Intel Cell Based Design Mentor Enviroment, vol 1. Santa Clara, Calif.: Intel Corporation, 1988.
Intel Introduction to Cell-Based Design. Santa Clara, Calif.: Intel Corporation, 1988.
Intel Components Quality/Reliabilty Handbook. Santa Clara, Calif.: Intel Corporation, 1989.
Intel Packaging Outline and Dimensions. Santa Clara, Calif.: Intel Corporation, 1988.
International Test Conference (1973-). *IEEE Computer Society Proceedings: International Test Conference.*
Iscroff, Ron: Selecting an ASIC vendor. *ASIC Technology and News.* vol. 1 no. 2 June 1989, pp. 1, 10-11.
Leonard, Milt: IC makers tackle complex radiation. *Electronics Design.* June 8,1989, p.51.
McAnney, William H., and Jacob Savir: *Built-in Test for VLSI: Pseudorandom Techniques.* New York : John Wiley & Sons, Inc., 1987.
McClean, Don, and Javier Romeu: Design for testability with JTAG test methods, *Electronics Design.* June 8, 1989, p.67.
McCluskey, Edward J.: Built in self-test structures, *IEEE Design and Test of Computers,* April 1985, p.29.
—: Built in self-test techniques, *IEEE Design and Test of Computers,* April 1985, p.21.
—: *Comparison of Test Pattern Generation Techniques.* Stanford University; Center for Reliable Computing.
—: Dictionary of common testability terms. *IEEE Design & Test of Computers,* June 1989, vol. 6 no. 3, p. 72.
—: *Logic Design Principles: With Emphasis on Testable Semicustom Circuits.* Englewood Cliffs, N.J.: Prentice-Hall, 1986.
McWilliams, Thomas Melvin: *Verification of Timing Constraints on Large Digital Systems.* Ann Arbor, Mich.: University Microfilms International, 1980.
Miczo, Alexander: *Digital Logic Testing and Simulation.* New York : Harper & Row, 1986.
Millman, Jacob, and Christos Halkias: *Integrated Electronics: Analog and Digital Circuits and System.* New York: McGraw-Hill Book Company, 1972.
Millman, Steven D., and Edward J. McCluskey: Detecting Bridging Faults with Stuck-at Test Sets. Stanford University; Computer Systems Laboratory.
Milnes, Arthur George: Semiconductor Devices and Integrated Electronics. New York: Van Nostrand Reinhold, 1980.
Muller, Richard S., and Theodore I. Kamins: *Device Electronics for Integrated Circuits.* John Wiley & Sons, Inc., New York, 1977.
Noyce, Robert N.: Microelectronics, *Scientific American,* vol. 237 no. 3, September 1977, p. 62.
Phister, Montgomery: *Logical Design of Digital Computers,* New York: John Wiley & Sons, Inc., 1958.
Pieper, Chris M.: Putting tester compatibility into your ASIC designs, *ASIC Technology and News,* vol. 1 no. 1., July 1989, pp. 10-15.
Pirsig, Robert M.: *Zen and the Art of Motorcycle Maintenance.* New York: Bantam Books, 1974.
Roth, Charles H.: *Fundamentals of Logic Design.* St. Paul, Minn.: West Publishing Company, 1979.
Runyon, Stan: Designers wait for tools to forge design-to-test links. *Electronic Engineering Times,* August 14, 1989, p.46.
Santoni, Andy: Focus report: Understanding test improves ASIC designs, *ASIC Technology and News,* vol. 1 No. 2, July 1989, pp. 28-29.
Stover, Allan C.: *Automatic Test Equipment.* New York: McGraw-Hill Book Company, 1984.
Sze, S. M.: *Physics of Semiconductor Devices,* 2d ed. New York: John Wiley & Sons, Inc., 1981.
—: *Semiconductor Devices, Physics and Technology.* New York: John Wiley & Sons, Inc., 1985.

Timoc, Constantin C.: *Selected Reprints on Logic Design for Testability*. Silver Spring, Md.: IEEE Computer Society Press, 1984.

Tsui F.: *LSI/VLSI Testability Design*. New York: McGraw-Hill Book Company, 1987.

Walker, Robert, et al.: A new look at an old ASIC dilemma: Cells vs. arrays, *ASIC Technology and News*, vol. 1 no. 1, June 1989, pp.14-18.

Wong, Thomas: BiCmos making inroads to CMOS, ECL markets, *ASIC Technology and News*, August 1989, p.29.

INDEX